RUSSIAN POLITICAL, ECONOMIC, AND SECURITY ISSUES

IMPLICATIONS OF CLIMATE CHANGE ON RUSSIA

Russian Political, Economic, and Security Issues

Additional books in this series can be found on Nova's website under the Series tab.

Additional E-books in this series can be found on Nova's website under the E-books tab.

Climate Change and its Causes, Effects and Prediction

Additional books in this series can be found on Nova's website under the Series tab.

Additional E-books in this series can be found on Nova's website under the E-books tab.

RUSSIAN POLITICAL, ECONOMIC, AND SECURITY ISSUES

IMPLICATIONS OF CLIMATE CHANGE ON RUSSIA

MICHELLE D. MONAHAN
AND
JONATHAN N. WILEY
EDITORS

Nova Science Publishers, Inc.
New York

For permission to use material from this book please contact us:
Telephone 631-231-7269; Fax 631-231-8175
Web Site: http://www.novapublishers.com

NOTICE TO THE READER

Additional color graphics may be available in the e-book version of this book.

LIBRARY OF CONGRESS CATALOGING-IN-PUBLICATION DATA
Implications of climate change on Russia / editors, Michelle D. Monahan and Jonathan N. Wiley.
p. cm.
Includes index.
ISBN 978-1-61122-851-9 (hardcover)
1. Climatic changes--Environmental aspects--Russia. 2. Climatic changes--Government policy--Russia. 3. Global warming--Government policy--Russia. I. Monahan, Michelle D. II. Wiley, Jonathan N.
QC903.2.R8I66 2010
363.738'740947--dc22
2010043902

Published by Nova Science Publishers, Inc. † New York

CONTENTS

Preface **vii**

Chapter 1 Russia: Impact of Climate Change to 2030 - A Commissioned Research Report **1**
National Intelligence Council

Chapter 2 Russia: The Impact of Climate Change to 2030 - Geopolitical Implications **57**
National Intelligence Council

Index **105**

PREFACE

The Russian Federation is already experiencing the impacts of climate change in the form of milder winters, melting permafrost, changing precipitation patterns, the spread of disease, and increased incidence of drought, flooding, and other extreme weather events. Many of these observed climate impacts are having concrete, negative effects on Russians' quality of life. By 2030, Russia will start to feel the impacts of climate change in relation to both water and food supply. This book identifies and summarizes the latest research related to the impact of climate change on Russia

Chapter 1- Russia is already experiencing the impacts of climate change in the form of milder winters; melting permafrost; changing precipitation patterns; the spread of disease; and increased incidence of drought, flooding, and other extreme weather events. Many of these observed climate impacts are having concrete, negative effects on Russians' quality of life. By 2030, Russia will start to feel the impacts of climate change in relation to both water and food supply. Nonetheless, a significant portion of the country's senior leaders continue to voice the view that a warming climate is a net benefit for Russia. Russia has a number of attributes that provide a greater capacity for resilience than some other industrialized countries and most developing countries. However, as the impacts of climate change continue and intensify over the coming years, Russia's capacity to adapt and protect its people will be severely tested.

Chapter 2- The National Intelligence Council-sponsored workshop entitled The Implications of Global Climate Change in Russia, held on May 18, 2009, brought together a panel of Russia experts to consider the probable effects of climate change on Russia from a social, political, and economic perspective. The panelists concluded that Russia is reaching a point where serious deterioration of its physical and human capital is a major obstacle to sustainable economic growth and Russia's capacity to adapt and protect its people will be tested out to 2030. Climate change will add additional stress to energy and transportation infrastructure burdens. However, given Russia's high overall state capacity and the mixed or comparatively tolerable nature of most anticipated climate change impacts over the next twenty years, climate change is unlikely to lead to a general failure of the Russian state.

In: Implications of Climate Change ... ISBN: 978-1-61122-851-9
Editor: Michelle D. Monahan et al.

Chapter 1

RUSSIA: IMPACT OF CLIMATE CHANGE TO 2030 - A COMMISSIONED RESEARCH REPORT*

National Intelligence Council

SCOPE NOTE

Following the publication in 2008 of the National Intelligence Assessment on the National Security Implications of Global Climate Change to 2030, the National Intelligence Council (NIC) embarked on a research effort to explore in greater detail the national security implications of climate change in six countries/regions of the world: India, China, Russia, North Africa, Mexico and the Caribbean, and Southeast Asia and the Pacific Island States. For each country/region we are adopting a three-phase approach.

- In the first phase, contracted research—such as this publication—explores the latestscientific findings on the impact of climate change in the specific region/country.
- In the second phase, a workshop or conference composed of experts from outside the Intelligence Community (IC) will determine if anticipated changes from the effects of climate change will force

* This is an edited, reformatted and augmented edition of a National Intelligence Council Special Report, Order Code NIC 2009-04D, dated April 2009.

inter- and intra-state migrations, cause economic hardship, or result in increased social tensions or state instability within the country/region.

- In the final phase, the NIC Long-Range Analysis Unit (LRAU) will lead an IC effort to identify and summarize for the policy community the anticipated impact on US national security.

EastLink Consulting, LLC, collaborating with the Joint Global Change Research Institute (JGCRI) and Battelle, Pacific Northwest Division (Battelle, PNWD), developed this assessment on the climate change impact on Russia through 2030 under a contract with SCITOR Corporation. The Central Intelligence Agency's Office of the Chief Scientist, serving as the Executive Agent for the DNI, supported and funded the contract.

This assessment identifies and summarizes the latest peer-reviewed research related to the impact of climate change on Russia, drawing on both the literature summarized in the latest Intergovernmental Panel on Climate Change (IPCC) assessment reports and on other peer-reviewed research literature and relevant reporting. It includes such impact as sea level rise, water availability, agricultural shifts, ecological disruptions and species extinctions, infrastructure at risk from extreme weather events (severity and frequency), and disease patterns. This paper addresses the extent to which regions within Russia are vulnerable to climate change impact. The targeted time frame is to 2030, although various studies referenced in this chapter have diverse time frames.

This assessment also identifies (Annex B) deficiencies in climate change data that would enhance the IC understanding of potential impact on Russia and other countries/regions.

Executive Summary

Russia is already experiencing the impacts of climate change in the form of milder winters; melting permafrost; changing precipitation patterns; the spread of disease; and increased incidence of drought, flooding, and other extreme weather events. Many of these observed climate impacts are having concrete, negative effects on Russians' quality of life. By 2030, Russia will start to feel the impacts of climate change in relation to both water and food supply. Nonetheless, a significant portion of the country's senior leaders continue to voice the view that a warming climate is a net benefit for Russia.

Russia has a number of attributes that provide a greater capacity for resilience than some other industrialized countries and most developing countries. However, as the impacts of climate change continue and intensify over the coming years, Russia's capacity to adapt and protect its people will be severely tested.

The most important impacts of climate change in Russia will likely include the following:

- Energy. A warming climate holds the possibility of milder and shorter heating seasons, which in turn may lead to reduced Russian energy demand. Increased water availability—particularly along those Siberian rivers that are used for hydroelectric power—should result in increased power production in certain parts of the country. However, existing and future energy infrastructure for the all-important petroleum industry will experience more pronounced challenges—structural subsidence, risks associated with river crossings, and construction difficulties as permafrost thaws earlier and deeper, impeding the construction of vital new production areas. These latter challenges have the potential for a material, negative impact on the single-greatest source of revenue to the Russian state—the oil and gas industry.
- **Water.** Many parts of Russia's massive territory will experience increases in the availability of water, including much of Siberia, the Far North, and northwestern Russia. This change will bring certain positive impacts—including for hydroelectric generation (above). However, managing the increased flows will pose other problems, especially when these increased flows coincide with extreme weather events such as downpours, or springtime ice-clogged floods. In addition, increasing water shortages are predicted for southern parts of European Russia, areas that already experience significant socioeconomic and sociopolitical stresses. Moreover, a number of densely populated Russian regions that are already subject to water shortages are expected to face even more pronounced difficulties in decades to come.
- **Agriculture.** As growing seasons become longer and precipitation patterns change, using lands for agricultural purposes that previously would have been too far north—too cold for too much of the year—will become possible. Raising new crops and new varieties of crops that are currently grown in Russia also could become possible.

However, a changing climate may not be hospitable to expanded agriculture. A key question is whether the longer growing seasons and the warmer Russian agricultural lands will result in increased yields. Yields of existing crops may fail and whether new crops will succeed remains to be seen. **Agriculture will become more reliant on irrigation (especially in the southern parts of European Russia), pesticides and herbicides, and more vulnerable to droughts and other extreme weather.**

- **Migration. Russia, which is already the number two destination for immigrants (after the United States) is likely to experience greater migration pressure from Central Asia, the Caucasus countries, Mongolia, and northeastern China.** These latter areas are expected to experience increased water shortages and resulting economic stress. In addition, internal migration pressures may occur as residents in Russia's many northern cities face increasing economic and climate-related challenges.
- **Accentuation of existing socioeconomic and sociopolitical stresses. Russia is better equipped to deal with the impacts of climate change than many of its neighbors. Nonetheless, by 2030, climate change appears likely to accentuate some of the stresses that currently plague Russia.** Some of the most affected regions are areas where already socioeconomic and sociopolitical relations are attenuated and unsettled. Most of the impacts of climate change will manifest themselves in smaller cities and in the Russian countryside. For example, the long- turbulent North Caucasus region will be drier, hotter, and less prosperous than it is today. The Primorskiy Kray and the Russian Far East, which have long struggled to develop peacefully next to China, appear likely to experience even greater migration pressures, which could exacerbate longstanding cross-border tensions.

INTRODUCTION AND BACKGROUND

Current Climatology of Russia[i]

Russia has the largest amount of land area of any country in the world. Most of this area is more than 400 kilometers from the sea, with the center of the country being almost 4,000 kilometers from the sea. The terrain ranges from grassy steppes in the south to frigid tundra in the polar north. The treeless, marshy tundra comprises almost 10 percent of the country. Russia's topography includes the world's deepest lake and Europe's highest mountain, and its landscape contains all the major vegetation zones of the world except a tropical rain forest. More than half of the country is above 60° north latitude and is covered with snow for almost half of the year.

Less than one percent of Russia's population lives in the northernmost part of the country, from the Finnish border to the Bering Strait. This area, shaped by glaciation in the last ice age, continues to be subject to erosion by frost weathering. Rivers here flow north to the Arctic Ocean, often hampering drainage of lakes and ponds across the tundra. Summer nights, called "white" nights, are so short that dawn comes shortly after dusk. Vegetation above the permafrost consists mostly of mosses, lichen, and dwarf trees and shrubs.

Russia's large forested region, called the taiga, comprises an area about the size of the United States and contains primarily coniferous trees such as spruce, cedar, larch, and fir. The region includes most of European Russia, and about one-third of Russia's people live there. The annual average temperature of this region is below freezing; the northern part of this region is one of the coldest inhabited areas on Earth.

The steppes, often imaged as typical Russian landscape, are treeless, grassy plains occasionally interrupted by mountain ranges. Located from south of Moscow to the Black and Caspian seas, this is the only region that has a relatively temperate climate and is suited to agriculture. However, the region occasionally experiences catastrophic droughts and short, intense periods of precipitation. At the southernmost part of the region, a narrow subtropical climate warms the edges of the Black Sea and provides Russia's only warm resort area.

Most of Russia receives little precipitation. In the south and east, mountain ranges prevent Indian and Pacific Ocean winds from bringing precipitation and warmer temperatures inland. The highest levels of precipitation are in the northwest region of the country, with levels decreasing toward southeast and European Russia. The wettest areas are along the Pacific coast and near the Caucasus. A monsoonal climate along Russia's Pacific coast brings seasonally high amounts of precipitation, reversing the direction of winds in summer and winter.

In winter, steady winds tend to blow from the south and southwest across most of the country. In summer, winds come from the north and northwest. This reversal of the winds causes less temperature variation than might be expected between winter and summer. For January, the average temperatures are -8 degrees Celsius (°C) in St. Petersburg, -27°C in the West Siberian Plain, and -43°C at Yakutsk (east-central Siberia, at about the same latitude as St. Petersburg). In the summer, the Arctic islands average 4°C, and the southernmost regions of Russia average 20°C.

PROJECTED REGIONAL CLIMATE CHANGE

Climate Observations

Temperature trends over most of the Arctic and northern Russia before about 1920 were likely dominated by natural variability.[ii] It is difficult to explain increasing temperatures since 1920 without including the impacts of human emissions of greenhouse gases. Average temperatures over the past decade are the warmest ever measured in the documented history of climate records in Russia. Studies by Roshydromet,[iii] the Federal Service for Hydrometeorology and Environmental Monitoring, show that annual average temperatures over Russia have increased significantly during the past 10 years; the models suggest a continuation of this trend over the next five to 10 years. Such conclusions are supported by the findings of other Russian agencies, the Russian Academy of Sciences in particular, and by most foreign scientists (Figure 1).

Data collected by the Roshydromet surface network of hydro-meteorological observations show that during 1990-2000 the mean annual surface air temperature increased by 0.4°C. (During the previous hundred years, the increase was only about 1.0°C.) Warming is more evident in winter and spring and more intensive east of the Urals.[iv]

Temperatures in the Arctic are rising at almost double the rate of the global average. In many inland Arctic regions, surface air temperatures have warmed 0.2°C per decade over the past 30 years. Sea ice in the Arctic has decreased by 3 percent per decade between 1978 and 1996, and summer sea ice thickness has decreased by 40 percent since the 1950s.[v] Precipitation at high latitudes has increased by 15 percent over the past decade, with most of this increase occurring over the past 40 years.[vi] Arctic summers are now warmer than at any time in at least the past 400 years.

The fourth Intergovernmental Panel on Climate Change (IPCC) climate change assessment (AR4)[vii] reported that in areas of the boreal north, the liquid precipitation season has become longer by up to three weeks over the past 50 years. Increasing winter temperatures in the northern regions has considerably changed the ice regime[1] of the region's water bodies. Comparing the years 2010-2015 with 1950-1979, the assessment predicts that, in the later period, ice cover duration on the rivers in Siberia is expected to be 15-27 days shorter and maximum ice cover thinner by 20-40 percent. Also, an annual increase of 5 percent was observed in river flow, with a winter increase of 25-90 percent over the base flow due to increased melt and thawing permafrost.

Winter snowfall and snow depth in the Northern Hemisphere's high latitude regions have increased during the past few decades; this trend is likely to be associated with increasing precipitation related to surface air warming. This trend is supported by significant positive trends in winter temperatures across much of the former Soviet Union in the past 50 years.[viii]

Most recent research shows that Siberian permafrost temperatures rose considerably during the latter half of the 20th century, although the extent to which this can be attributed entirely to climate warming is currently unknown. Recent research revealed positive warming trends for all permafrost regions in response to positive trends in air temperature, with the strongest warming trend in regions of continuous permafrost. A slight cooling trend is found only for the topmost soil layers in regions of seasonally frozen ground at the southern margins of the region draining into the Arctic.

Melting permafrost serves as another revealing indicator of climate change. Significant areas of the Russian permafrost zone, which covers 60 percent of the country (the largest such region in the world falling under a single nation's jurisdiction), clearly show a trend of temperature increase in the top layers of frozen ground from the 1970s to the 1990s, corresponding with the warming of the atmosphere. Although climate change in European Russia is less severe than in Siberia, the change in the condition of frozen terrain is no less substantial. In the past 20-30 years, temperatures in the frozen ground of Russia's European Arctic and Subarctic have increased between 0.22 and 1.56°C, matching increases in the number and thickness of taliks (thawed underground pockets). These observations suggest a progressive increase in seasonally thawing soil, as well as a 14-80 percent increase in thawed pockets of soil in individual regions of the Russian Arctic.[ix]

Areas of seasonal frost have also shifted noticeably northward, and the area of isolated and sporadic pockets of frozen soil has decreased.[x] Although deeper layers of frozen soil are insulated against thawing by icy strata and

[1] An ice regime is a region of generally consistent ice conditions. Ice types are measured in a range from grey ice (0-. 15 m) to permanent ice (>3 m). See ftp://ftp2.chc.nrc.ca/CRTreports/ISOPE_98_IRS

organic soil and vegetation, models suggest that deeper seasonal thawing may change the composition of plant and animal communities.[xi] Natural tundra will likely grow smaller disappear entirely as a result.[2] [xii]

Satellite-derived measurements of snowfall show a spring and summertime decrease, likely due to increased temperatures. Snow accumulation over Russia accounts for about 5 percent of fresh water discharge to the Arctic Ocean. Significant changes in fresh water discharge have affected the salinity, sea ice distribution and circulation of the Arctic and nearby oceans. North of 50° N latitude, annual precipitation has increased by about 4 percent over the past 50 years, especially over Russia's permafrost-free zone and the entire Great Russian Plain. Over northern Russia, snow is providing a declining fraction of total annual precipitation.

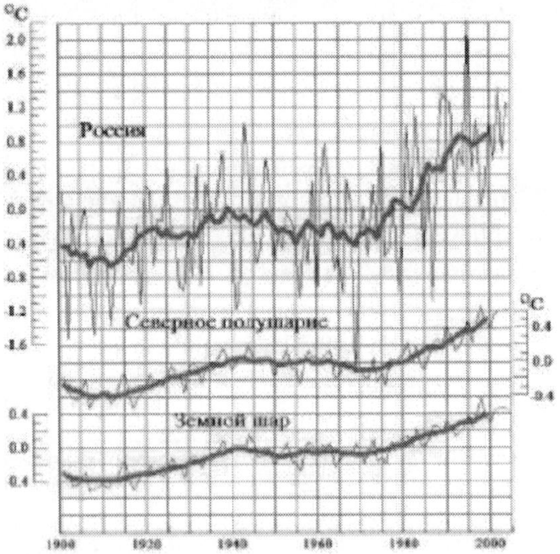

Figure 1. Surface air temperature increase in Russia, the Northern Hemisphere, and the world, 1900-2004. Source: Dobrolyubova, Julia, *Climate Change Effects and Assessment of Adaptation Potential in the Russian Federation* (Moscow: Russian Regional Environmental Centre, November 19-20, 2007), slides.

---

[2] The source does not provide an exact date of when natural tundra depletion may occur, other than mentioning that the change in composition of plant and animal communities is already occurring: "Although deeper layers of frozen soil are insulated against thawing by intermediate icy strata and a layer of organic soil and vegetation, models demonstrate that further deepening of seasonal thawing as a result of rising air temperatures may upset that balance. Should this happen, it will change (and this is already occurring) the composition of plant and animal communities, and existing natural complexes of the tundra may severely dwindle, or disappear entirely." http://assets

The 20th century saw a trend of increased river output from the six largest Eurasian rivers flowing into the Arctic Ocean.[xiii] A similar trend is found in the climate simulation for the same period by the Hadley Centre's coupled climate model when the effects of manmade greenhouse gases are included. This finding is in line with predictions that global warming will cause changes in the water cycle.

Studies have shown that runoff in the Lena River increases in winter, spring, and (especially) the summer; and discharges decrease in autumn. These changes in seasonal streamflow characteristics indicate a hydrologic regime shift toward early snowmelt and higher summer streamflow, perhaps due to regional climate warming and permafrost degradation in the southern parts of Siberia. Winter snow accumulation is a major influence on summer and autumn discharge of the Ob and Yenisey Rivers and can affect winter and spring discharges of the Lena River, suggesting the importance of topography and permafrost conditions to river discharges in high-latitude regions.

Climate Predictions (Modeling)

Although Global Circulation (or Climate) Models (GCMs) can be used to infer climate changes in specific regions, developing models that have a high resolution sufficient to resolve local and regional scale changes is preferable. There are many challenges in reliably simulating and attributing observed temperature changes at regional and local scales. At these scales, natural climate variability can be relatively larger, making it harder to distinguish long-term changes expected due to external forcings.

The procedure of estimating the response at local scales based on results predicted at larger scales is known as “downscaling.” The two main methods for deriving information about the local climate are (1) dynamical downscaling (also referred to as “nested modeling” using “regional climate models” or “limited area models”), and (2) statistical downscaling (also referred to as “empirical” or “statistical-empirical” downscaling). Chemical composition models include the emission of gases and particles as inputs and simulate their chemical interactions; global transport by winds; and removal by rain, snow, and deposition to the earth’s surface.

Downscaled regional climate models rely on global models to provide boundary conditions and the radiative effect of well-mixed greenhouse gases for the region to be modeled. There are three primary approaches to numerical downscaling: (1) limited-area models, (2) stretched-grid models, and (3) uniformly high resolution atmospheric GCMs (AGCMs) or coupled atmosphere-ocean (-sea ice) GCMs (AOGCMs).

The magnitudes and patterns of the projected rainfall changes differ significantly among models, probably due to their coarse resolution. The Atlantic and Pacific Oceans are strongly influenced by natural variability occurring on decadal scales, but the Indian Ocean appears to be exhibiting a steady warming. Natural variability (from ENSO, for example) in ocean-atmosphere dynamics can lead to important differences in regional rates of surface-ocean warming that affect the atmospheric circulation and hence warming over land surfaces. Including sulfate aerosols in the models damps the regional climate sensitivity, but greenhouse warming still dominates the changes. Models that include emissions of short-lived radiatively active gases and particles suggest that future climate changes could significantly increase maximum ozone levels in already polluted regions. Projected growth of emissions of radiatively active gases and particles in the models suggest that they may significantly influence the climate, even out to year 2100.

Stabilization emissions scenarios assume future emissions based on an internally consistent set of assumptions about driving forces (such as population, socioeconomic development, and technological change) and their key relationships. These emissions are constrained so that the resulting atmospheric concentrations of the substance level off at a predetermined value in the future. For example, if one assumes the global CO_2 concentrations are stabilized at 450 parts per million (ppm) (the current value is about 380 ppm), the climate models can be tuned to produce this result. The tuned model predictions for regional climate changes can be used to assess specific impacts at this stabilization level. A more detailed discussion of the ability of the models to project regional climate changes can be found in Annex A.

Climate Projections of Future Temperature and Precipitation

The IPCC AR4 does not include predictions specifically focused on Russia. Most of the country is included within a modeling region referred to as Northern Asia (NAS). Warming in this region is expected to be well above the global mean—consistent with the more general result that high latitudes will (and are) warming more than low latitudes (see Figure 2). This warming is particularly great in autumn and early winter when sea ice is thinnest and the snow depth is insufficient to blur the relationship between surface air temperature and sea ice thickness. Precipitation is also very likely to increase (Annex A).[xiv]

The Arctic is extremely vulnerable to climate change.[xv] The region is warming much more rapidly than the global average. The IPCC report states that the winter warming of northern high latitude regions by the end of the century will be at least 40 percent greater than the global mean, based on a number of models and emissions scenarios. Temperature increases for the central Arctic are projected to be about 3-4°C during the next 50 years. Even an optimistic scenario for projecting future greenhouse gas emissions yields a result of a 4°C increase in autumn and winter average temperatures in the Arctic by the end of this century. Recent satellite data show that the area covered with perennial ice in the Arctic Ocean has receded significantly in recent years, falling to nearly half the area observed in 2005.

During the 21st century, the thaw depth will increase substantially, summer soil moisture will eventually be reduced, and a poleward movement of the permafrost extent is expected. Based on three global climate models (Canadian Climate Center scenario, GFDL scenario and ECHAM scenario), a 3 0-40 percent increase in active layer thickness for most of the permafrost area is projected, with the largest relative increases concentrated in the northernmost locations.[xvi]

Regionally, the changes are a response to both increased temperature and increased precipitation (changes in circulation patterns). In a few regions, Siberia for example, the amount of snow is projected to increase because of the increase in precipitation (snowfall) from autumn to winter. Consistent results from the majority of the current generation of models show, for a future warmer climate, that a poleward shift of storm tracks occurs with greater storm activity at higher latitudes.[xvii]

Most models ignore the effect of land cover change in future projections. Past and future changes in land cover may affect the climate in several ways, causing changes in albedo, in the ratio of latent to sensible heat, and therefore in surface temperature, and in CO_2 fluxes to and from the land. No coupled AOGCM has included all the effects of land cover changes. The general consensus is that land cover changes may be very important at the regional level, where these changes occur.

At the regional level, the Russian Federation has considerable experience in climate modeling, with three centers of research: St.Petersburg V.A. Fock Institute of Physics; Institute for Numerical Mathematics (INM) in Moscow, and the Oboukhov Institute of Atmospheric Physics in the Russian Academy of Sciences (IAP-RAS) in Moscow. Both the INM and the IAP-RAS have their own climate models, although only the former submitted simulation data as part of the IPCC fourth assessment process.

Table 1 shows the 2080-2099 temperature and precipitation projections from a set of 21 global models. The numbers represent average (mean) changes from the period 1980-1999 over the 21 models. For each season, the minimum, 25 percent, 50 percent, 75 percent, and maximum changes are shown. For example, for the winter (DJF, or December, January and February), the average minimum temperature increase is 2.9°C, and the average precipitation change is +12 percent.

The five-to-10-year projections of the Russian hydro-dynamical climate models[xviii] match very well with the model projections of the IPCC AR4[xix], when the same scenarios and assumptions are used.[xx] These projections suggest that the mean annual surface air temperature over Russia will increase over the next five to 10 years by 0.60°C±0.2 from the annual mean temperature in the year 2000. The increase in temperature will vary by region, but by 2015 the average winter temperatures will have increased by an additional 1°C. In summer, the increase is only expected to be 0.40°C. During this same period, annual averaged precipitation is projected to increase by 4-6 percent, with the increase being as much as 7-9 percent north of Eastern Siberia.[xxi]

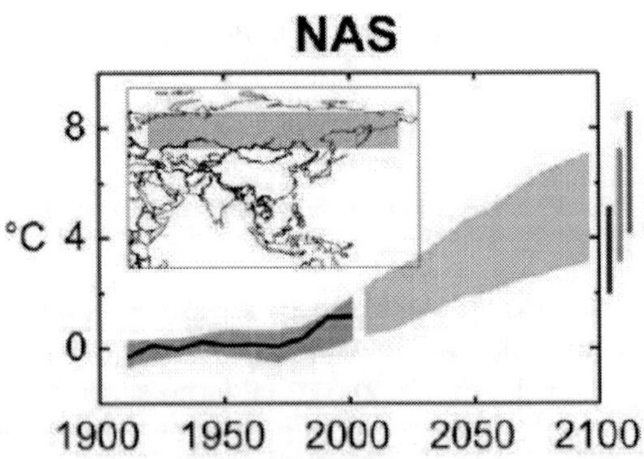

Figure 2. IPCC projected temperature increases for Northern Asia (NAS) (including Russia). Temperature anomalies with respect to 1901-1950 for the NAS land region for 1906-2005 (black line) and as simulated (red envelope) by multi-model datasets incorporating known forcings; and as projected for 2001-2 100 for the A1B scenario (orange envelope). The bars at right represent the range of projected changes for 2091-2100 for the B1 (blue), the A1B (orange) and the A2 (red) scenarios. Source: Climate Change Risk Management Ltd, "Climate Change in Russia: research and impacts" (May 2008), hyperlink (accessed February 17, 2009): http://www.uk-russia-ccproject.info/documents/Impacts in Russia Report 2008.pdf

Table 1. Regional Averages of Temperature and Precipitation Projections from a Set of 21 Global Models in the MMD for the A1B Scenario for the Region of Northern Asia (NAS: 50°N, 40°E: 70°N, 1 80°E)

	Temperature Response					Precipitation Response				
Season	Min	25	50	75	Max	Min	25	50	75	Max
DJF	2.9	4.8	6.0	6.6	8.7	12	20	26	37	55
MAM	2.0	2.9	3.7	5.0	6.8	2	16	18	24	26
JJA	2.0	2.7	3.0	4.9	5.6	-1	6	9	12	16
SON	2.8	3.6	4.8	5.8	6.9	7	15	17	19	29
Annual	2.7	3.4	4.3	5.3	6.4	10	12	15	19	25

This table is UNCLASSIFIED.

The table shows the minimum, maximum, median (50 percent), and 25 and 75 percent quartile values among the 21 models, for temperature (°C) and precipitation (percent) change. The changes are calculated as the 2080-2099 mean with respect to the 1980-1999 mean. DJF=December January February, MAM= March April May, JJA=June July August, SON=September October November. Source: Table 11.1 in IPCC [Intergovernmental Panel on Climate Change]. *Climate Change 2007: the Physical Science Basis*, ed. S. Solomon, D. Qin, M. Manning, M. Marquis, K. Averyt, M.M.B. Tignor, H.L. Jr. Miller, and Z. Chen (Cambridge: Cambridge University Press, 2007).

Maximum temperature increases are expected to occur in the winter in the Arctic.[xxii] By the middle of the 21st century, temperatures are projected to rise as much as 4-5°C in the Arkhangelsk region, the Komi Republic, the Yamalo-Nenets Autonomous Area, and over Taimyr.[xxiii] Temperature increases in the summer in these regions are small. However, in the southern regions, such as in the Northern Caucasus, the Volga region, and in the south of Western Siberia, an increase of 2–3°C is projected.

According to an assessment done by the World Wildlife Fund (2008),[xxiv] an appreciable increase in winter precipitation totals is expected by 2050—notably, a 30 percent increase on the Taymyr Peninsula and a 15-20 percent increase in Chukotka and the Barents Sea region. This increase in precipitation is expected to continue throughout the second half of the century. Total precipitation will more than double current values in the eastern Russian Arctic, consequently forming a deep layer of snow and reducing the period of soil freeze in winter. Alternatively, summer precipitation totals will increase only 5-10 percent by 2050, and 10-20 percent by the end of the 21st century, with the increase being slightly larger in the eastern part of the Arctic. An increase in the frequency of heavy rainfall is forecasted for the same region, effectively accelerating coastline erosion. Throughout the Arctic, there will be more rainfall than evaporation, despite predicted increases in evaporation due to warming.[xxv] The result is the formation of bogs[3], more likely prominent along the central and eastern Arctic coast.

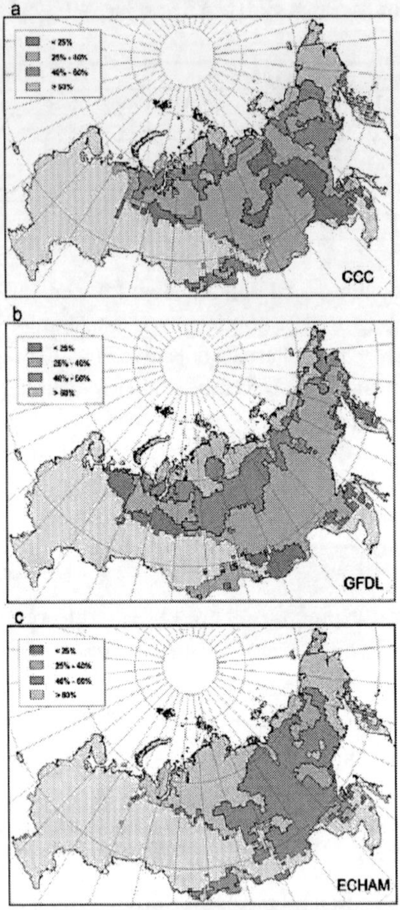

Figure 3. Projected 2050 changes of the active-layer permafrost thickness in northern Eurasia, relative to present-day simulations, based on forcing from three different global climate models: (a) CCC (Canadian Climate Center) scenario; (b) GFDL scenario; (c) ECHAM scenario (From Anisimov and Reneva 2006). Source: Climate Change Risk Management Ltd, "Climate Change in Russia: research and impacts" (May 2008), http://www.uk-russia-ccproject.info/documents/Impacts_in_Russia_Report_2008.pdf (accessed February 17, 2009).

---

[3] According to the US Environmental Protection Agency, bogs are "characterized by spongy peat deposits, acidic waters, and a floor covered by a thick carpet of sphagnum moss. Bogs receive all or most of their water from precipitation rather than from runoff, groundwater or streams. As a result, bogs are low in the nutrients needed for plant growth, a condition that is enhanced by acid forming peat mosses." See http://www.epa.gov/owow/wetlands

Trends of wintertime snow mass accumulation vary over the country. In European Russia (that is, Russia east of the Urals) and south of Western Siberia snow mass is expected to decrease compared with long-term mean values. By 2015 a 10-15 percent decrease is expected. In most of the rest of Russia, snow accumulation is expected to increase by 2-4 percent.[xxvi]

Projected changes in annual river runoff vary across the country. Winter runoff is projected to increase from 60-90 percent in the Central and Volga Federal Districts, and from 5-40 percent in other Federal Districts. In the Black Earth area and in the south of the Siberia Federal District, springtime river runoff is projected to decrease by 10-20 percent.[xxvii]

Permafrost and Arctic Ice Projections

According to a collaborative report led by Climate Change Risk Management (CCRM),[xxviii] seasonal thaw depths are predicted to increase around 2050 by more than 50 percent in the northernmost permafrost regions, and 30-50 percent elsewhere. By 2100, it is predicted that almost 60 percent of current permafrost regions will thaw and freeze on a seasonal basis.[xxix] Increased precipitation contributes to the thawing of frozen soils and is projected to lead to a 14 percent increase in freshwater discharge into the Arctic Ocean. Modelers warn that there are significant uncertainties in model projections of changes to the permafrost.[xxx]

A coupled climate-permafrost model was used by Anisimov and Renava (2006)[xxxi] to calculate changes in permafrost extent and thickness for three timeslices. Model results predict a reduction of near-surface permafrost area by 11 percent, 18 percent, and 23 percent by 2030, 2050, and 2080, respectively. Contractions of near-surface permafrost over these same periods are 18 percent, 29 percent, and 41 percent, respectively.

Despite the uncertainties, most modelers agree that seasonal thaw depths will increase by more than 50 percent in the northern Russia, including much of Siberia and the Far East; and by 30 percent to 50 percent in most other permafrost regions.[xxxii] Figure 3 shows the projected changes in active-layer permafrost thickness in northern Eurasia by 2050. Increased methane emissions from the melting permafrost will be a significant feedback on radiative forcing and climate change. Projected changes in the permafrost to the year 2050 are shown in Figure 3.

Most models project that summer ice will decline much more rapidly than winter ice.[xxxiii] Arctic sea ice is projected to decrease more rapidly than other sea ice.[xxx iv] Some scientists suggest that the Arctic Ocean could be ice free in summer in the next 10-20 years.[xxxv]

Projections of Changes in Agricultural Growing Seasons

The decline in the number of very cold winters in many regions across Russia has led to better conditions for growing winter crops. In the Central Black-Earth and Volga regions, the frequency of very cold winters has decreased from an average of 18-22 percent in the period up to 1990 to 8-10 percent in the past several years.[xxxvi] In Northern Caucasia, this frequency has been reduced from 10 percent to 4 percent.

Conditions for growing corn have improved in many areas of European Russia. In the Stavropol Territory, "climate-related"[xxxvii] corn yield has increased 30 percent over the past 20 years, but in parts of Asian Russia (e.g., the Baikal), corn yield has decreased.

From 1970 to 2000, the growing season (with air temperatures above +5°C) lengthened by an average of approximately 5-10 days over much of the agricultural region in European Russia. However, frost-free periods did not lengthen.[xxxviii]

If this trend continues, agricultural production may increase significantly by 2015. The growing season is likely to be significantly lengthened. Both the growing season and duration of frost-free days may be increased on the order of 10-20 days per year.[xxxix] Many plant species may experience a northward migration of growing boundaries. On the Siberian rivers and in the Kama River basin, a reduction of the freeze period of as many as 15-27 days is expected by 2010-2015.

Changes in the Frequency or Strength of Extreme Climatic Events

Changes in the frequency of extreme events may be one of the most damaging consequences of climate change. Climate change over the past 10-20 years in Russia has been linked to extreme events, including heat waves, floods, and fires. The IPCC assessment[xl] reports a substantial increase in the number of days with more than 10 millimeters of rain in Siberia, causing a 50-70 percent increase in surface runoff. There were also a significant increase in the number of fires in Siberian peatlands and more frequent flooding in Russian Arctic rivers due to heavy rain and earlier breakup of river ice.[xli] Satellite measurements show that vegetation fires, mostly forest fires, occurred over about 10 million hectares during 1997-2003. Outbreaks of disease-carrying insects also occurred in the northern part of the country, where outbreaks have never been observed in the past.[xlii]

Observations suggest that large floods are already more frequent.[xliii] By 2015, there is likely to be more flooding in river basins in the Archangelsk Region, the Komi Republic, the Ural area, and in the basins of Enisei and Lena.[xliv] In the Arctic, increased water discharges occurring earlier in the spring may be blocked by ice jams, causing the duration of inundated flood plains to increase from the current 12 days to 24 days. In the past five years, the Lena, one of the world's 10 largest rivers, has experienced two floods more severe that any previous recorded flood.[xlv]

Ice-jam-induced floods in the Lena River Basin are expected to double by 2015. Flooding in the Far East and the Maritime areas is expected to double or triple. In the mountain and submountain regions of Northern Caucasia (Republics of North Caucasia, Stavropol Region) and in the Western and Eastern Sayan Mountains, more mudflow and landslide hazards are expected.[xlvi] In St. Petersburg, the probability of a disastrous flood is expected to increase in the next 5-10 years.[xlvii]

The average annual discharge of fresh water from the six largest Eurasian rivers to the Arctic Ocean increased by 7 percent between 1936 and 1999.[xlviii] Peterson et al. (2002), Wu et al. (2005) and Shiklomanov et al. (2006)[xlix] project Russian river discharges will continue to increase at an accelerated rate. The average projected change in annual discharge in large Russian rivers is around 15 percent (range: -12 to 45 percent).[l] Annual discharge in the Yenisey, Ob, Lena and Kolyma rivers are projected to change by 6-45 percent, -12-45 percent, 12-45 percent, and 10-45 percent, respectively.[li] All experienced significantly larger increases in winter discharges—as much as 325 percent in a high-sensitivity scenario (4°C).[lii]

Over this century, increases in the frequency and intensity of heat waves are expected in western and central Europe, possibly including parts of Russia. A record-breaking heat wave occurred in central Europe in summer 2003. This event was the hottest since instrumental records began around 1780 (1.4°C above the previous warmest in 1807) and is very likely to have been the hottest since at least 1500.[liii]

Hazardous events due to the changes in permafrost are expected to increase by 2015.[liv] Melting of permafrost islands will lead to increases in landslides, mudflows, and other dramatic and abrupt changes in the landscape When a glacier recedes, unstable glacial lakes are formed that increase the likelihood of glacier-related outbursts and debris slides. Glacier retreat between 1985 and 2000 has resulted in a 3-6 percent increase in the proportion of glaciers covered by debris, increasing the melt rate and the likelihood of glacier-related significant events such as rock and mud slides.[lvi]

On both the Baltic and Pacific coasts, a rise in sea level may result in the coastline being more vulnerable to tsunamis. Studies of the Baltic region have stressed the possibility that tsunami activity could profoundly affect the coastline. Some of the largest tsunamis ever observed have occurred along the Pacific coast, which is prone to tsunamis. The IPCC 2007 assessment identifies the Baltic and White seas as areas of probable increased flooding and erosion.[lvii]

Over large areas of Russia, the number of both high-intensity and mid-intensity fire-hazard days is expected to increase. By 2015 the number of fire-hazard days may increase by more than five days in a season on most of the territory. The areas most likely to experience an increased duration of fire-hazard days (more than 7 days in a season) include areas south of the Khanty-Mansi Autonomous Area; and in Kurgan, Omsk, Novosibirsk, Kemerovo and Tomsk Regions, Krasnoyarsk and Altai Territories, Sakha-Yakutia Republic.[lviii]

IMPACTS OF CLIMATE CHANGE ON HUMAN-NATURAL SYSTEMS

In Russia, the socioeconomic impact of climate change has long been controversial. Some of Russia's most prominent climate scientists have argued persistently that a warming climate will bring net positive benefits for a cold, massive country whose territory includes vast expanses of permafrost and undeveloped forests, while others posit some unmitigated negatives. While the debate continues among Russian observers, the weight of scientific evidence points to a more complicated picture—some significant benefits, as well as profound problems for human systems that have the potential to challenge Russia's ability to respond.

Economic Growth and Development

The Russia that will face the unfolding impacts of climate change between now and 2030 is a Russia newly grown accustomed to relative wealth. No longer is it the economic basket case that it was in the 1 980s and 1 990s. In fact, until the global financial crisis began to bite in Russia in late 2008, the country had ridden a decade-long wave of economic good fortune following the crash of the Russian ruble in August 1998.

Russia is extraordinarily dependent on its extractive industries and commodity production. In addition, the country's development has been highly uneven, with most wealth concentrated in the capital, Moscow. Only a modest trickle-down effect has occurred in smaller cities, and virtually none has occurred in rural areas. Most of the impacts of climate change will manifest themselves in smaller cities and in the Russian countryside.

Another consideration that relates to the advance of climate change impacts is the role of the government in people's everyday lives. During Soviet times, government was highly intrusive but simultaneously was the source of considerable private skepticism. ("We pretend to work; they pretend to pay us," was one of the core folk wisdoms.) Government did, however, provide services that ensured a minimum standard of living for nearly all citizens. In today's Russia, many people have arguably even less expectation that the government will provide for their minimum requirements. But if climate change begins to wreak serious humanitarian impacts, such as recurrent massive flooding or the collapse of aging infrastructure, and if the government is not in a position to respond in a commensurate way, one of the key questions will be whether climate change prompts political unrest. For the time being, we judge this to be a significant, but unresolvable open question.

Energy Systems

The stable operation of energy systems is a major technological challenge for a country as massive as Russia. It is also a matter of vital importance to everyday Russians, whose day-to-day survival depends on the timely availability of heat and power in the face of Russia's severe climate. Energy systems are no less of a matter of survival for the Russian economy, which stays afloat largely due to petroleum exports. For these reasons, the impacts of climate change on Russian energy systems are of exceptionally great importance.

The seasonality and geographical scope of climate change across Russia have significant implications for Russian changing energy demand between now and 2030. As is mentioned above, mean temperature increase in Russia by 2030 is projected to be significant nationwide, although more pronounced in the north and east of the country than in the south and west, and more pronounced in winter and spring than in the summer or fall.[lix]

As shown in Figure 4, the projected climate warming is expected to lead to a reduction in the length and intensity of the heating season, which has the potential to result in reduced energy consumption for heating. Roshydromet projects that on average, by 2015, the heating season will be three to five days shorter across the entire country.[lx] In the eastern regions of Primorskiy Kray, Sakhalin, and Kamchatka, the heating season may be more than five days shorter by 2015. Some regions may experience little if any reduction in the length of the heating season. These trends are projected to extend and intensify by 2030 and beyond.

In areas that experience this change, residents and workers may experience greater indoor comfort, as heating systems and building envelopes will be better able to cope with the heating load.

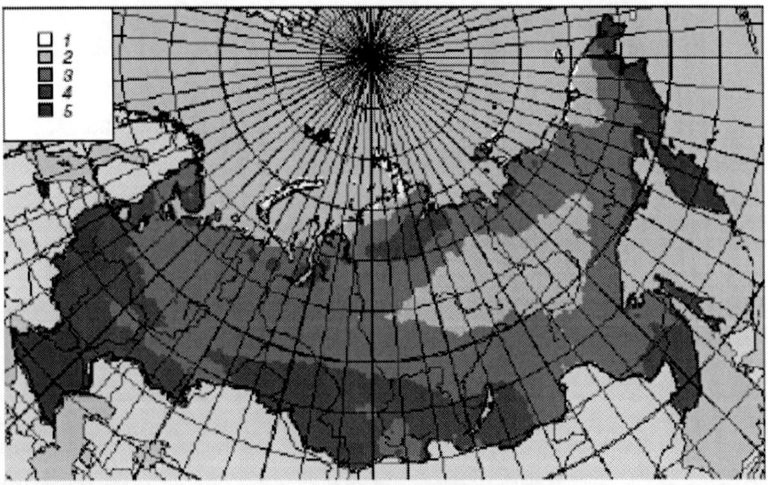

Figure 4. Reduction in duration of the heating season. Legend: Zone 1 shows *reduction* of the length of the heating season by 0.0-1.9 percent, Zone 2 -- 2.0-3.9 percent, Zone 3 -- 4.0-5.9 percent, Zone 4 -- 6.0-7.9 percent, Zone 5 -- 8.0-10.0 percent. Source: Federal Service for Hydrometeorology and Environmental Monitoring (Roshydromet), "Assessment Report on Climate Change and its Consequences in Russian Federation" (Moscow, 2008)

For those regions that do experience a reduction, the extent of the related energy savings is a matter of some debate, even in official Russian government projections. According to analysis presented in 2005 and 2008 by the Russian Federal Service on Hydrometeorology (Roshydromet), the reduction in heating days resulting from a warmer climate may not translate into saved fuel. Even if there are fewer total heating days, they may stretch out over the same period of the year–or even longer[lxi] –because of increased temperature variability.[lxii]

This Roshydromet analysis contrasts with the Russian Federation's Fourth National Communication under the UN Framework Convention on Climate Change, which was submitted in 2006. According to the Fourth National Communication, the reduction in heating requirements by 2025 will result in a net fuel savings of 5-10 percent nationwide (and greenhouse gas emissions reduction of 2 percent).[lxiii] If the latter projection proves accurate, the saved fuel could provide a significant economic benefit, and potentially a significant balance-of-payments benefit, provided that the saved fuel was exported instead of being consumed domestically.

Just as a changing climate is expected to affect energy demand in Russia, so too will it affect energy supply. On the supply side of the ledger, the changing climate may affect hydroelectric power production, electricity transmission and distribution systems, and petroleum production and transportation systems.

Hydroelectric power production will realize some benefit, and some negative impacts, associated with the increase in flows of rivers that are used for hydroelectric production. As is true with the question of the length of the heating season, the net benefit or cost in 2020 remains ambiguous.

Many of the major Russian rivers will experience increased water flows due to glacial melt and selective regional precipitation changes. For the most part, this change will offer opportunities for increased power production. According to Roshydromet, the Volga-Kamsk Cascade will experience a net increase of 10-20 percent in water flows. The reservoirs throughout the Northwest Federal District will experience a 5-10 percent increase, and the massive Siberian power dams along the Angarsk-Yenisey, Vilyu, Kolyma, and Zeya will experience increases ranging as high as 15 percent. In addition, certain hydro-electric reservoirs in the southern part of the country will experience reductions in productivity due to reduced water flow. Nonetheless, operating regimes for all power dams will require review in light of anticipated climate change, according to Roshydromet. In addition, there will be increased challenges related to managing head and tail waters in the face of increased flows, and particularly in relation to increased incidence of extreme downpours.[lxiv]

Another energy supply-related impact from climate change to 2030 will relate to electricity transmission systems. One form of the heightened risk to power transmission will come from permafrost melt and the resulting creation of thermokarst and other unstable soil conditions. High- voltage power lines will be one of the many kinds of structures that will be susceptible to damage as upper soil layers thaw and re-freeze. One particularly vulnerable transmission system will be the lines serving the Bilibino nuclear power plant on the Arctic coast and running from the town of Chersk to Pevek.

Another heightened risk for power transmission systems will be increased wind load on power stanchions, as on other large structures. Power lines in the North Caucasus, as well as in the regions of Murmansk, Arkhangelsk, Leningrad, portions of Sakha (Yakutia), Irkutsk, Magadan, Khantiy-Mansiysk, and Evenkia will be exposed to 20 percent increases in wind force and may need to be reconstructed or reinforced as a consequence.[lxv]

Climate change by 2030 appears unlikely to sufficiently affect the Russian electric power sector as to lead to significant national security implications for the Russians. However, major power system failures could lead to serious human hardship and could therefore conceivably fuel political dissatisfaction in Russia. If a major portion of one of Russia's regional power grids were destabilized by the failure of a major power dam, or if power supply lines failed due to unusually abrupt winds in the North Caucasus, for example, one could envision the potential for localized instability.

If the hydropower and power transmission industries face challenges as climate change intensifies, even greater challenges face the petroleum industry, with likely greater significance for the Russian economy and state. In today's Russia, oil and gas are the predominant components of economic performance. Together they represent on the order of 60 percent of total exports and one-third or more of state revenues.[lxvi] Russia learned, starting in late 2008, that its economy was therefore at risk of significant volatility in case of a downturn in global energy prices, and diversifying the economy away from extractive industries is a stated goal of the Russian government. However, for the foreseeable future, the core of Russia's economy will remain oil and gas.[lxvii]

This in turn means that Russia's economy is highly vulnerable to climate change impacts that affect the current or future operations of the petroleum sector. Many areas that are currently the focus of exploration and production activity will be more difficult to exploit. Pipeline and rail transportation systems that cross major rivers and permafrost will be subjected to unprecedented stresses and strains, many of which were not anticipated when initial design parameters were established. Critical new upstream development areas, such as the Yamal Peninsula, will be more complicated to reach by land and harder to develop in the face of thawing permafrost and shorter winter seasons.

The Russian petroleum industry has traditionally centered in West Siberia and the Volga region, with transportation links extending to European portions of Russia and then to western and central European markets. The Russian gas industry has centered on three super-giant fields in the Nadym-Pur-Taz region—the Urengoy, Yamburg, and Medvezh'ye fields. At present, in addition to thousands of producing oil and gas wells, Russia has roughly 50,000 kilometers of oil pipelines and roughly 150,000 kilometers of gas pipelines, most of which were constructed in the 1980s under Soviet rule. There are also scores of processing plants and refineries distributed across Russia's massive territory.[lxviii] (See Figure 5 for a map of existing and planned pipelines and gas production regions.)

The core climate-related vulnerability facing oil and gas pipeline systems is that these systems were designed and built with the presumption of a stable climate. The thousands of river crossings did not provide margins of error to accommodate the increased water flow that will result from climate change by 2030. They were not constructed using horizontal directional drilling techniques that allow deeper and more secure passage under riverbeds. Underwater river crossings in several key producing and transit regions are thought to be particularly at risk—the upper and lower Volga and its tributaries in the regions of Nizhegorodskaya, Orenburg, Perm, Samara, Saratov, Ulyanovsk, Bashkortostan, Tatarstan, Tyumen, Novosibirsk, and Sakhalin among others.[lxix]

In addition to climate-related risks for river crossings, oil and gas pipelines and other facilities are at risk in permafrost regions. In these areas, pipelines and other structures are typically constructed above ground to allow thermal insulation to avoid thawing the soil. In the period to 2030, however, these regions will experience deeper seasonal thawing, resulting in structural subsidence and weakened integrity of pipelines and other petroleum-industry installations.

The permafrost zones are also exceptionally important for the future development of oil and especially gas production. Russian gas production has been the basis of not only Russian export earnings but also Russia's controversial, growing politico-economic power vis-à-vis central and western

European neighbors, as was demonstrated again in early 2009 during the Russian- Ukrainian gas crisis.

Maintaining Russian gas exports is therefore a matter of highest national priority. There has been a dramatic decline of production at the Urengoy, Yamburg, and Medvezh'ye fields that have been the core of Russian gas production since the end of the Soviet period.[lxx] New production is crucial for Gazprom to realize its production targets and satisfy both domestic requirements and export consumers in coming years.[lxxi]

Three of the key areas that the Russians expect to produce new gas will keenly feel the effects of a changing climate by 2030. First, the Yamal peninsula is an Arctic region that is a vast wealth of untapped gas prospects. According to some Gazprom projections, it could account for as much as 200 billion cubic meters (bcm) of gas production per year by 2020, and 360 bcm per year by 2030.[lxxii] However, developing Yamal will be significantly complicated by a changing climate. Supplies that will need to be brought in by land will require the construction of new roads and rail links, which will be tricky with the growth of thermokarst. Previous techniques, like the use of seasonal ice roads will be more problematic due to the shorter cold season. New above-ground pipelines and other elevated installations will have to be constructed using deeper foundations to avoid structural damage from subsidence.

A second key area for new Russian gas production is the Barents Sea. Here, one massive new field called Shtokman is to be developed some 550 kilometers north of the Kola Peninsula, with a projected annual production of around 90 bcm of gas.[lxxiii] This hugely challenging technical undertaking, which will require the construction of ice-capable production platforms in more than 300 meters of water, is especially difficult because it is so far offshore that it is beyond the range of helicopters, yet it is vulnerable to seasonal pack ice and vicious storms. In the face of a rapidly changing Arctic climate, vessels traveling to Shtokman will have to navigate increasingly severe waters and endure bitter winter storms.

A third key projected area for the Russian gas industry is Eastern Siberia and the Russian Far East. Here too, climate change in the period to 2030 will pose increasing complications—melting permafrost, swollen rivers, more frequent and severe storms and more prevalent incidence of traditionally atypical forms of disease. As mentioned above, the Russian gas industry has traditionally been oriented toward customers within Russia and in neighboring European countries. However, in the past decade, Gazprom and the Russian government have identified the goal of moving to the East and developing gas resources that can feed to the Pacific Rim. The official Eastern Gas Program released in August 2007 projects total extraction of 100 bcm/year by 2030.[lxxiv] New fields are being developed in the Sea of Okhotsk, near Sakhalin Island. Other prospects are being pursued as far west as the area to the north of Lake Baikal (e.g., the Kovykta field) and in the Sakha Republic (Yakutia). All of these projects will require major new construction with countless major and minor river crossings and a significant number of permafrost operations. This development will therefore be vulnerable to the same kinds of challenges from climate impacts as have been discussed above.

Food Production and Drinking Water Supply

Russia will experience a mix of positive and negative impacts on food and water supply in the period to 2030. The net impacts in these important areas will depend heavily on the extent to which adaptation measures can be implemented in an affordable and timely manner, but doing so will be difficult.

Experts project that Russia will experience an increase in total water supply in the period to 2030. According to Roshydromet, in the aggregate Russia will experience an 8-10 percent increase in water volume by 201 5–the equivalent of a 12-14 percent increase per capita–with these trends expected to continue in the years that follow.

That said, different regions will experience significantly different changes in their respective water supply. The northern and northwestern portions of European Russia, as well as the central Volga, many of the non-Chernozem lands, the Urals, and the Russian Far East will experience increasing water availability. In the dams along the VolgaKamsk Cascade, water flows are projected to increase by 10-20 percent by 2015, as mentioned above. In the Northwest federal district, dams will see a 5-10 percent increase over the same period. And some of the key Siberian rivers systems–the AngarskYenisey, the Vilyu, the Kolyma, and the Zeya–will experience flow increases by up to 15 percent.[lxxv]

On the other hand, many other parts of Russia will experience worsening water shortages, including densely populated industrial regions that are projected to experience increases in water demand of 5-25 percent.[lxxvi] In the Chernozem lands, these water-poor areas will include the Belgorod, Voronezh, Kursk, Lipetsk, Orel, and Tambov regions. In the south, Kalmykia, Krasnodar, Stavropol, and Rostov regions will face increasingly challenging water situations, with reductions in water supply on the order of 5-15 percent.[lxxvii] In southwestern Siberia, the list will include Altay, Kemerovo, Novosibirsk, Omsk, and Tomsk. Across a key southern belt, a whole host of Russian regions will face mounting, and serious, water problems. Included in this list will be both certain key agricultural lands (more on food supply below), and also a number of key industrial regions. Even the capital and the Moscow Oblast' will face "particularly acute" water supply problems.[lxxviii]

Regarding food supply, the longstanding popular presumption in Russia has been that a warmer global climate would translate into a significantly more hospitable Russian environment for agricultural production. Indeed, there are several respects in which climate change by 2030 will reduce longstanding challenges for Russian agriculture. First and foremost, growing seasons have already become longer and are predicted to become longer still.[lxxix]

Accompanying this change will be a reduction in the frequency of winter temperatures that are sufficiently bitter to damage winter plantings. More sensitive varieties of winter plantings will be possible in much of Russia by 2030, and it will be possible to plant existing varieties farther north than would have been the case in the past.[lxxx] For example, it will be possible to plant longer-ripening grains and late-ripening sugar beets as far north as Moscow.[lxxxi] Interestingly, the longer growing seasons will not be accompanied by an increased frost-free period except in the Northwestern, Central, and Volga federal districts.[lxxxii]

Based on temperature ranges expected by 2030, it will also be possible to introduce entirely new crops that are not widely grown in Russia today. For example, the projected temperature of the north Caucasus and the lower Volga will be well suited to intensive agriculture for crops that are typically found in Central Asia and the south Caucasus at present–crops such as cotton, grapes, tea, citrus, and other fruits and vegetables.[lxxxiii]

A key question, however, is whether the longer growing seasons and the warmer Russian agricultural lands will result in increased yields. In fact, this does not appear to be assured—at least not based on the crops that are currently raised. Many of the current "bread basket" areas of Russia–including the Black Earth or Chernozem lands, the lower Volga region, and the southern part of Siberia–will experience reductions in grain yields resulting from reduced precipitation–reductions in yields of more than 22 percent by 2020. Warmer average temperatures will produce better grain yields in some parts of the country that have not traditionally served as the heartland for grain production. Regions such as the Northwest and Central federal districts and the Volga-Vyatsk region are expected to see a 10-15 percent increment in grain yields. Nationwide, according to Roshydromet, grain yields could shrink by more than 11 percent by 2020.[lxxxiv]

Plant diseases and pests will become a more serious challenge in many parts of Russia. In the southern part of European Russia and in western Siberia, locusts are expected to be increasingly common. Already they are found more frequently than was the case two or three decades ago, and they are expected to be even more prevalent in the future in the Stavropol, Kalmykia, Volgograd, Astrakhan, Saratov, and Rostov regions, and in some parts of southern Siberia.[lxxxv] In northwestern Russia, farmers are experiencing an infestation of Colorado potato beetles, which are now found in Karelia. In coming years, as mild winters become increasingly common, they are expected to spread into southern parts of the Arkhangelsk region and the Komi Republic.[lxxxvi]

A third question that arises about future agriculture is whether human management and distribution systems, and rural society itself, will be able to adapt in a timely manner to manage new crops, new supply chains, and requirements. Indeed, rural Russia has typically been resistant to change. In addition, supply, distribution, and management issues have historically posed great hurdles for Russian agriculture. A key question will be whether a true national market for food and agricultural products develops, or whether Russian regions persist in semi-national, semi-intra-region forms of agricultural trade.[lxxxvii]

Additional challenges for agriculture by 2030 will come from the increased frequency of severe weather events. Periods of drought in key agricultural regions are expected to be 50-100 percent more frequent by 2015, with the trend line continuing thereafter.[lxxxviii]

By 2030, Russia will start to feel the impacts of climate change in relation to both water and food supply. To maintain stable food supply, significant changes will be required in terms of varieties that are planted, the lands that are used for agriculture, and the extent and intensity of pesticide and irrigation use. All of these solutions are theoretically possible, but none will come easily or inexpensively. All will test the ability of Russian authorities and Russian agriculture to adapt quickly as climate change impacts are felt.

Transportation Systems

Transportation systems are another aspect of Russia's socioeconomic life that will experience major impacts from climate change by 2030. For the most part, these impacts will entail the need for significant adaptations, which will imply significant capital requirements. This will be true for Russia's extensive rail networks as well as its more limited road networks. However, in relation to river transportation and especially Far North maritime transport, a changing climate will open new and likely beneficial possibilities.

Russia's railways are the backbone of its goods and passenger transportation system, with over 87,000 kilometers of railroads stretching across most of the country.[lxxxix] The system is a state-owned monopoly that moves 1.3 billion passengers and 1.3 billion tons of freight annually, which represents 83 percent of all freight in Russia (not counting petroleum pipeline operations).[xc] The rail system is most highly developed in the European part of Russia and along the southern reaches of Siberia. Nonetheless, a significant portion of the rail system–such as the Baikal-Amur Mainline (BAM)–cross permafrost zones, so they will be subject to the same risks of subsidence and structural weakening from permafrost melt as well as increased vulnerability at river crossings from increased Siberian river flows, as discussed above.

Russia's road transport systems are much less significant for cross-country transportation, but they too will be affected by climate change. Some of the impacts may be positive, in that reduced winter snowpack in European Russia (where the road system is much more developed and much more used than in the Arctic or the Far East) may reduce road hazards and wear and tear on existing roads. Other impacts will be negative, however. The increase in weather variability, with elevated risks of severe storms and downpours, may lead to elevated dangers for road transportation and risks of mudslides and erosion in mountainous areas and near rivers and floodplains. In the Far North and Far East, where wintertime ice roads have been a means of wintertime survival, shorter cold seasons will result in significantly reduced road transport capacity.[xci]

River transport, which is another key element of Russia's total transportation system, will experience both new problems and benefits from climate change to 2030, varying by region. In areas such as the Don River Basin in Russia's southwest, where there will be a reduction in total water flow, river navigation may encounter serious challenges. Extensive, and expensive, dredging may be required to allow continued barge and river freighter traffic.

In other areas, where river flows will significantly increase, such as along the major Siberian river systems and in the northwest of the country, river transport may be enhanced and facilitated. The exception to this rule may occur in areas where the earlier arrival of spring weather and the more pronounced melt-off of winter snow and ice lead to river ice jams and flooding.[xcii]

Another potentially significant transportation impact from climate change is the increased possibility of sea passage through Arctic waters. The so-called Northern Sea Route (NSR) that runs from near the island of Novaya Zemlya to the Bering Strait offers the prospect of up to a 40 percent savings in sea distance for journeys between northern Europe and Pacific Rim ports in either North America or northeastern Asia.[xciii] By 2020, the navigation season along the NSR will increase from around 36 days at present to around 40 days per summer.[xciv] Furthermore, the reduction in the extent of Arctic ice will allow vessels to travel in deeper waters farther from shore. Nonetheless, Arctic Sea shipping will not be without its share of challenges. Icebergs will continue to pose hazards to navigation, and bitter storms may produce significant wave action.

Human Health

Climate change may present Russia with a host of new and unwelcome challenges by 2030—both in the form of dangers related directly to climate and in the form of pest- borne disease.

In the case of direct effects, the combination of more frequent droughts and heat waves has had an impact on vulnerable populations in Russia already.[xcv] By 2030, as extreme weather events become more prevalent, this kind of increased risk to human health will rise further, particularly affecting the aged and infirm, especially for those unable to afford residential air conditioning.

Historically, Russia's bitter winters served as a check on the populations of many disease-carrying pests. Rodent populations, mosquitoes, and ticks were limited by the rigor of the seasons.

In recent years, however, these historical factors have receded. For example, according to one report, Russia's current rodent population is ten times higher than historical norms. Worse yet, one-third of the rodent population is estimated to carry one of the viruses that cause hemorrhagic fever with renal syndrome (or HFRS), which is a deadly illness if not caught early in its course.[xcvi] Incidences of HFRS spike after each occurrence of an unusually mild Russian winter.

As in other northern countries, mosquitoes have always been a summertime challenge in Russia. (Due to the poor quality of the housing stock, mosquitoes were often an unexpected wintertime challenge too; they would live through the winter in standing water in Russian basements, as many Western students and diplomats experienced.) But by 2030, they are expected to pose an increasing public health threat. As may as 250,000 Russians suffer from latent, local forms of malaria. West Nile and Denge Fever are reported to be spreading across the country as well.[xcvii]

Ticks are another disease vector that will grow worse by 2030. Tick encephalitis, Lyme disease, and tick rickettsiosis (Rocky Mountain Spotted Fever) are three of the diseases that are spreading increasingly aggressively across Russia.

Intestinal diseases are also a risk for Russia in the period to 2030. This risk will be especially significant in southern European Russia and the northern Caucasus region, where fresh water supply and water quality are expected to deteriorate as a result of climate change.[xcviii] However, even in distant Yakutia, in 2002, early spring flooding, which will also be increasingly common in Siberia by 2030, triggered a massive outbreak of enteric fever.[xcix]

Coping Capabilities in Facing Natural Disasters

Russia is better equipped than many other countries to respond to disasters resulting from climate change, certainly much better equipped than most of its regional neighbors. The central entity involved in governmental response to natural and manmade disasters is the Russian Federation Ministry of Civil Defense, Emergency Situations, and Disaster Response (known in Russia by the shorthand "Ministry of Emergency Situations," and often referred to in the West as "Emercom"). This organization brings together many of the functions that fall under the US Department of Homeland Security, including the Federal Emergency Management Agency (FEMA) and the Coast Guard, as well as local fire departments all across the country.[c] It is a proud, well-recognized organization that has earned public respect for its involvement in responding to a number of tragic occurrences in recent years.

Of particular significance for response to climate change impacts will be the Ministry's units with responsibility for forest fire prevention and response, maritime emergencies, flood protection and response, and search and rescue. The Ministry is Moscow-based but has regional centers across Russia, including several in southern Siberia that could be especially important in ensuring timely response to climate change-related disasters.

Another important component of Russia's coping capacity comes from the Russian Federal Service for Hydrometeorology and Environmental Monitoring (Roshydromet or Hydromet). Hydromet is the rough equivalent of the US National Oceanic and Atmospheric Administration (NOAA) and is active in monitoring, assessment, analysis, and prediction of weather and climate. Hydromet operates the Russian weather service, including over 1,600 meteorological stations across Russia,[ci] as well as serving as the leading scientific organization and the lead Russian representative for international negotiations and scientific undertakings related to climate change. Its research institutes work in close collaboration with institutes under the Russian Academy of Sciences, as well as leading Russian participation in the IPCC and other scientific assessment and forecasting activities.

Despite the considerable capabilities of both of these governmental organizations to analyze and respond to natural disasters, Russia will face a number of challenges in this context. Effective adaptation to climate change will require the application of huge resources and, more difficult, careful policy reforms. For example, policymakers will have to decide what to do with the residents of Russia's outsized northern urban centers. These Arctic cities are a monument to the sensibilities of Soviet planners—and an economic disaster.[cii] Thawing permafrost will pose more problems for these cities, and many of these cities sit on or near the banks of the Siberian rivers that will experience significant increases in flows and increased risks of flooding. Yet mass relocation would be both costly and politically challenging.

Other Urban Infrastructure

In the period to 2030, climate change could have a variety of impacts on urban infrastructure in Russia. Some of these impacts have already been discussed above, such as the potential for a reduction in heating requirements and heating loads that could accompany an increase in wintertime temperatures.

Another broad category of impact—but a negative one—is projected sea-level rise. This impact has the potential to bring significant challenges to a host of Russian cities and port complexes. Particularly vulnerable is Russia's second city, St. Petersburg, which is already regularly at risk of flooding when strong winds blow to the east from the Gulf of Finland. This vulnerability will only rise as sea level rises and storm surges grow more intense. [ciii] The risks of catastrophic flooding in St. Petersburg before 2030, and of consequent damage to both the economy and to unique historical buildings, is great.

St. Petersburg is not the only city at risk. The level of the Black Sea has been rising since the 1 920s, and the rate of rise has increased significantly since the 1 980s (currently about 2 centimeters per year).[civ] This will affect Russia's main warm water port complex at Novorossiysk, where dry cargoes, crude oil, and refined petroleum products are all exported. It will also affect Russia's main Black Sea military base, which is at Sevastopol, in neighboring Ukraine. Outside the Black Sea area, Russia's vital deepwater Atlantic basin port at Murmansk, which comprises both military and civilian capacity, will also be at risk to rising sea levels, as will the Pacific Rim ports, including Vladivostok and others.

As a general matter, Russia's population is projected to experience significant risks in the period to 2030 and beyond from extreme weather events—floods, torrential rains, severe winds, tornados, hurricanes, and the like. Officially designated dangerous hydrometeorological events across all of Russia have been growing markedly more common for the past decade-plus.[cv] This trend is expected to continue to 2030 and afterward. Russia's major urban centers may experience periods of drought combined with heat waves. In such circumstances, the risks of heat-related or disease-related illness could rise significantly.

International Issues

The international treatment of the Arctic over the next 20 or so years and questions of immigration related to climate change will affect Russia.

Greenhouse warming may bring greater changes in the Arctic than anywhere else on the planet. As has been discussed above, pronounced warming is already eroding the polar ice cap, and the thermal qualities of open water are contributing in turn to further warming and further melt-off.

This all translates into the Arctic being a much less imposing and more hospitable place than it has been in the past. Summers will bring increasingly extensive open seas that will facilitate speedy sea transportation of goods between northern Europe and the Pacific coasts of North America and northeastern Asia.[cvi] (See transportation discussion above.) New ports are being built along the Russian Arctic coast.

In addition, the warmer Arctic is engendering increased interest on the part of all of the littoral states in off-shore development. In 2007, the Russian polar scientist and politician Artur Chilingarov led an undersea expedition intended to bolster Russian claims that the Arctic is predominantly within the exclusive economic zone of Russia. One of Chilingarov' s key arguments is that the undersea Lomonosov Ridge extends from Russian territory and therefore validates Russia's claim to half of Arctic Ocean. Complicating matters further, there is no clear agreement as to the legal regime that should govern competing claims in the Arctic.[cvii] Chilingarov's expedition culminated with the depositing of a Russian flag on the sea bottom, some 2.5 miles below the surface. "The Arctic has special geopolitical importance for Russia," Chilingarov later said.[cviii]

Given that Russia is not alone in its strong economic and security interests in the Arctic, climate changes that affect the Arctic could prompt the development of new military bases and activity. In late January 2009, military and political leaders from NATO met in Reykjavik, Iceland, to discuss how to manage the opportunities and challenges posed by a warming Arctic. The Secretary-General of NATO, Jaap de Hoop Scheffer, told the assembled audience, "Climate change is not a fanciful idea. It is already a reality, a reality that brings with it certain new challenges, including for NATO. Several Arctic rim countries are strengthening their capabilities, and military activity in the High North region has been steadily increasing."[cix] It is thus possible to imagine a significant increase in military presence in the Arctic, beyond what has been the case since the end of the Cold War.

Another international issue facing Russia is climate-related migration. Already today, Russia is the world's second biggest destination for migration (after the United States), attracting an estimated seven million migrants in 2008, of whom only about four million were legal.[cx] At present, most migrants present in Russia are from the countries of Central Asia and the Caucasus, and they seek economic opportunity to help support families in their countries of origin. Many migrants are involved in construction, other manual labor, and trading, especially in foodstuffs. In the Russian Far East and Primorskiy Kray, there are also a significant number of temporary workers from northeastern China. In short, today's migration to Russia seems to be significantly motivated by the "pull" phenomenon of economic opportunity.

By 2030, migration may become more of a "push" phenomenon. Water availability is projected to become an increasingly serious challenge in Central Asia, Mongolia, and northeastern China, and simultaneously droughts are projected to become more frequent. Glacial-fed rivers are at risk of becoming more and more depleted by 2030.[cxi] The ability of the Central Asian states to adapt to a changing climate may well be more limited than is the case with

Russia. In turn, migration may become a source of instability within Russia, especially in difficult economic times. Already today, nationalist and reactionary political and social groupings are committing increasing numbers of hate crimes in Russian cities and towns. A Moscow-based nongovernmental organization that monitors hate crimes recorded over 500 attacks against foreigners in 2008, a one-third increase over 2007.[cxii]

Adaptive Capacity

The impacts of climate change will be felt differentially, depending upon how well a society can cope with or adapt to climate change, that is, its adaptive capacity. Adaptive capacity is defined by the IPCC as "The ability of a system to adjust to climate change (including climate variability and extremes) to moderate potential damages, to take advantage of opportunities, or to cope with the consequences."[cxiii] Although the specific determinants (or "drivers") of adaptive capacity are a matter of debate among researchers, there is good agreement that economic, human, and environmental resources are essential elements. Some components of this adaptive capacity are near term, such as the ability to deliver aid swiftly to those affected by, e.g., flooding or droughts. Other components include a high enough level of education so that people can change livelihoods, sufficient unmanaged land that can be brought into food production, and institutions that provide knowledge and assistance in times of change. For instance, Yohe and Tol[cxiv] identified eight qualitative "determinants of adaptive capacity," many of which are societal in character, although the scientists draw on an economic vocabulary and framing:

1. The range of available technological options for adaptation.
2. The availability of resources and their distribution across the population.
3. The structure of critical institutions, the derivative allocation of decision-making authority, and the decision criteria that would be employed.
4. The stock of human capital, including education and personal security.
5. The stock of social capital, including the definition of property rights.
6. The system's access to risk-spreading processes.
7. The ability of decision-makers to manage information, the processes by which these decision-makers determine which information is credible, and the credibility of the decision-makers themselves.

8. The public's perceived attribution of the source of stress and the significance of exposure to its local manifestations.

Russian Adaptive Capacity in a Global Context

Researchers have only recently taken on the challenge of assessing adaptive capacity in a comparative, quantitative framework. A comprehensive global comparative study[cxv] of resilience to climate change (including adaptive capacity) was conducted using the Vulnerability-Resilience Indicators Model (VRIM—see box below).

Adaptive capacity, as assessed in this study, consists of seven variables (in three sectors), chosen to represent societal characteristics important to a country's ability to cope with and adapt to climate change:

Human and Civic Resources

- *Dependency ratio:* proxy for social and economic resources available for adaptation after meeting basic needs.
- *Literacy:* proxy for human capital generally, especially the ability to adapt by changing employment.

Economic Capacity

- *GDP (market) per capita:* proxy for economic well-being in general, especially access to markets, technology, and other resources useful for adaptation.
- *Income equity:* proxy for the potential of all people in a country or state to participate in the economic benefits available.

Environmental Capacity

- *Percent of land that is unmanaged:* proxy for potential for economic use or increased crop productivity and for ecosystem health (e.g., ability of plants and animals to migrate under climate change).
- *Sulfur dioxide per unit land area:* proxy for air quality and, through sulfur deposition, other stresses on ecosystems.
- *Population density:* proxy for population pressures on ecosystems (e.g., adequate food production for a given population).

Adaptive capacity for a sample of 10 countries from the 160-country study is shown in Figure 6 (base year of 2005). There is a wide range of adaptive capacity represented by these countries. Russia ranks high, both in the sample and overall:

- Russia ranks 32nd and Libya 34th (in the highest quartile).
- Indonesia ranks 45th, Belize 48th, Mexico 59th, and China 75th (in the second quartile).
- The Philippines ranks 91st and India 119th (in the third quartile).
- Morocco ranks 136th and Haiti 156th (in the lowest quartile).

Any country-level analysis must take into account the comparative ranking of the country.

METHODOLOGICAL DESCRIPTION OF THE VULNERABILITY-RESILIENCE INDICATOR MODEL (VRIM)

The VRIM is a hierarchical model with four levels. The vulnerability index (level 1) is derived from two indicators (level 2): sensitivity (how systems could be negatively affected by climate change) and adaptive capacity (the capability of a society to maintain, minimize loss of, or maximize gains in welfare). Sensitivity and adaptive capacity, in turn, are composed of sectors (level 3). For adaptive capacity these sectors are human resources, economic capacity, and environmental capacity. For sensitivity, the sectors are settlement/infrastructure, food security, ecosystems, human health, and water resources. Each of these sectors is composed of one to three proxies (level 4). The proxies under adaptive capacity are as follows: human resource proxies are the dependency ratio and literacy rate; economic capacity proxies are GDP (market) per capita and income equity; and environmental capacity proxies are population density, sulfur dioxide divided by state area, and percent of unmanaged land. Proxies in the sensitivity sectors are water availability, fertilizer use per agricultural land area, percent of managed land, life expectancy, birth rate, protein demand, cereal production per agricultural land area, sanitation access, access to safe drinking water, and population at risk due to sea level rise.

Each of the hierarchical level values is composed of the geometric means of participating values. Proxy values are indexed by determining their location within the range of proxy values over all countries or states. The final calculation of resilience is the geometric mean of the adaptive capacity and sensitivity.

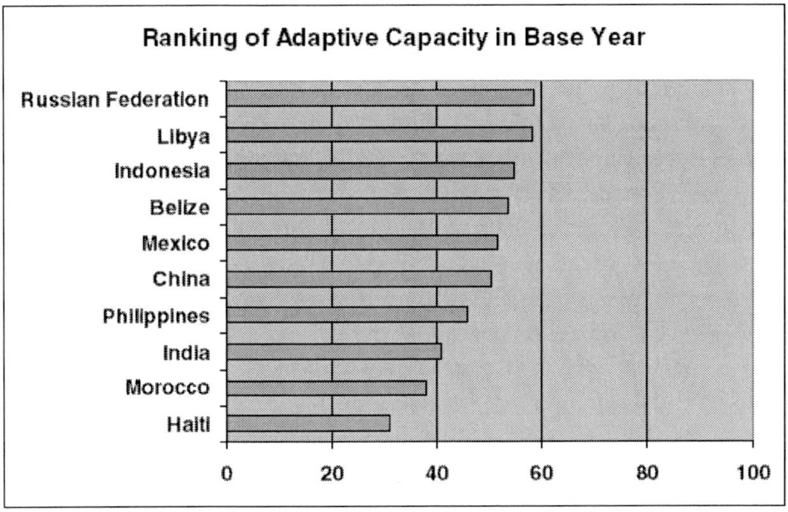

Figure 6. Sample of 10 countries' rankings of adaptive capacity (2005)

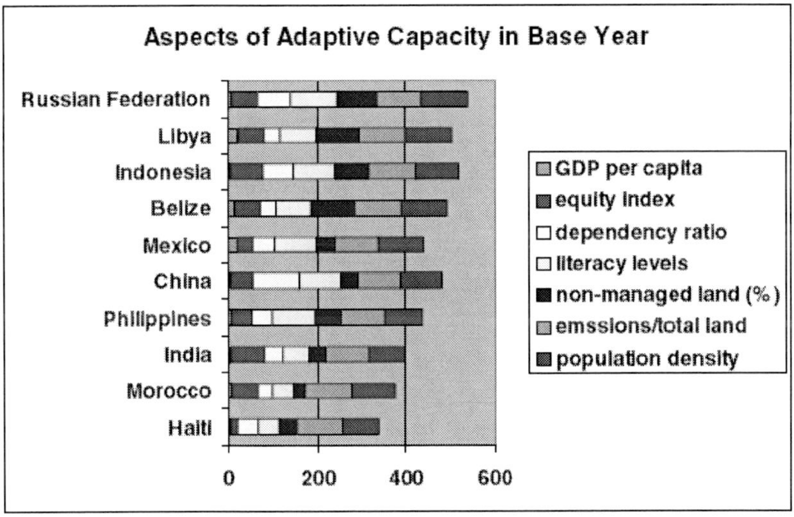

Figure 7. Variables' contributions to adaptive capacity rankings

Figure 7 shows the contribution of each variable to the overall ranking (slight differences occurring because of the methodology [see box]). In current adaptive capacity, Russia ranks first among the 10 countries shown in Figure 6. Russia's comparatively high literacy levels (indicating higher human capital), low greenhouse gas emissions, and low population density (indicating a less of a burden on the environment) more than compensate for low GDP per capita. This corresponds roughly with the pattern that one would expect for Russia. Russia has some significant areas of vulnerability, but it also stands to see some beneficial impacts from climate change. It has a well-educated populace, an economy that has some diversification but not a great deal, and a socioeconomic picture in which there is a small middle class, a small cluster of people with great wealth, and many who have only limited means.

Figure 8 shows projected adaptive capacity growth over time for the 10-country sample. Projections are made for two scenarios; rates of growth are based on the IPCC's A1 scenario in its *Special Report on Emissions Scenarios*.[cxvi] VRIM simulates two different hypothetical development tracks out to the year 2065 (well beyond the timescale of the present study) with intermediate results at 15-year time steps. These alternative development tracks are not intended to be predictive; they are scenarios.

Both scenarios feature moderate population growth and a tendency toward convergence in affluence (with market-based solutions, rapid technological progress, and improving human welfare). The scenarios used in this study differ in the rate of economic growth, one modeling high-and-fast economic growth, the other delayed growth.

Over time, a low-growth scenario widens the gap among the 10 countries—and the high- growth scenario widens the gap even more (Figure 8). In both scenarios, China's high economic growth (indicated by GDP per capita), favorable dependency ratio, and literacy rate allow that country to overtake Russia by 2050. In both scenarios, the strengths and weaknesses of current Russian adaptive capacity persist, with slow economic growth being a notable weakness.

Looking forward, Russia's ability to cope with climate change impacts to 2030 and beyond will obviously depend on both the nature and extent of the impacts and the extent to which Russia's adaptive capacities develop over time. This in turn depends on the nature and extent of Russia's socioeconomic and sociopolitical development over the coming years.

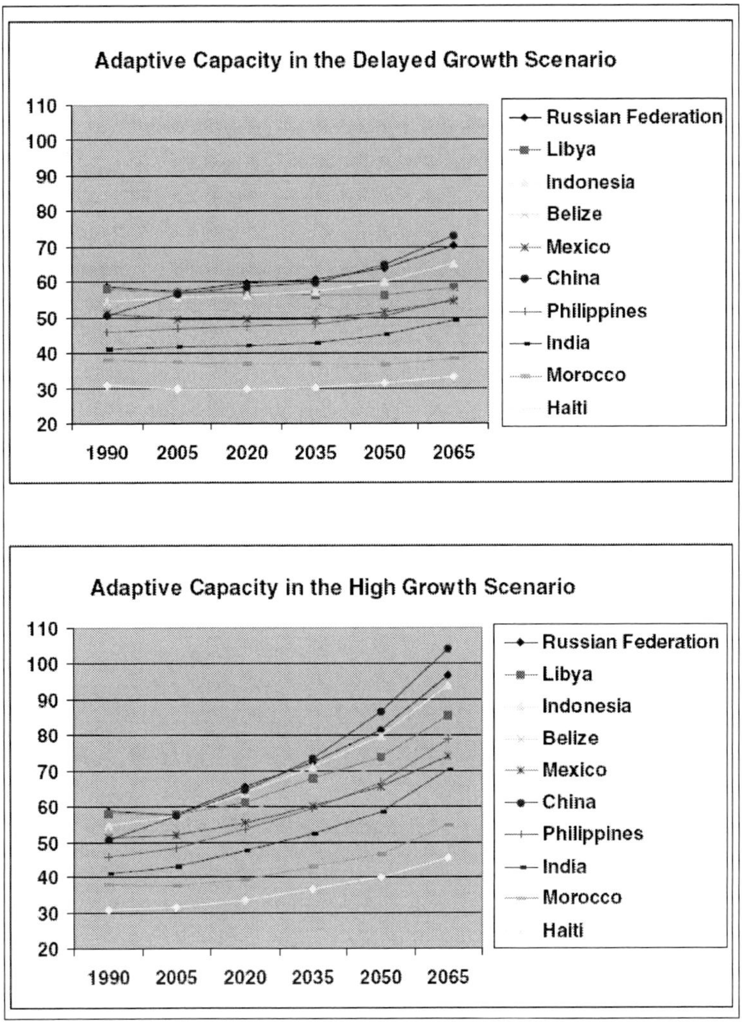

Figure 8. Projections of adaptive capacity for 10 countries

In the delayed-growth scenario, Russia's position is much less influenced by wealth accumulation (adaptive capacity) and much more heavily influenced by water availability, the production of cereal grains, and the use of fertilizer (included in impacts/sensitivity rather than in adaptive capacity). In this scenario, Russia's adaptive capacity still improves over time, but its progress, like that of other countries, is much more modest than in the high-growth scenario.

The high-growth scenario could, in the case of Russia, be consistent with robust early- period revenues from hydrocarbons and other commodity production, leading to significant wealth accumulation in the country and, over time, greater economic diversification of the sort that has been advocated in the last year by President Dmitriy Medvedev. (The elements and implications of this high-growth scenario are outwardly consistent with the track that Russia appeared to be on until the collapse of energy prices in the past six months.) In this scenario, Russia's adaptive capacity grows significantly over time, with Russia ending the period second only to China.

In the real world, a variety of factors will play into the overall calculus of Russia's adaptive capacity. Evolving socioeconomic conditions will be one key factor. To highlight this fact, one can compare the adaptation of the Russian oil industry to the adaption of its gas industry in the period since the breakup of the USSR. In the late 1990s and the first years of the current decade, the Russian oil industry experienced rapid innovation that reflected new ownership forms, new managerial techniques, and the introduction of international technology. Production skyrocketed, especially after the 1998 financial crash, while environmental impacts for the oil industry as a whole (spills, emissions, and accidents) dropped. By comparison, the relatively traditionalist Russian gas industry, where there was substantially less commercial, managerial, and technological change, evolved less dramatically.

In the area of Russian agriculture, socioeconomic forms and institutions most likely will be significant in determining the efficacy of the sector's adaptation to climate change.[cxvii] One particular shortcoming will be the relative weakness of agricultural education and training, akin to the extension service programs operated by the US Department of Agriculture. Russia's rural population is generally the country's most conservative social grouping; adapting to changing climatic conditions will require innovation that has historically been alien to much of rural Russian society.

Strengths/Weaknesses in Adaptive Capacity Assessments

Even comparative measures of adaptive capacity only allow analysts to ask better focused questions about area or local conditions that contribute to or reduce resilience. It is likely, for instance, that for particular places in Russia important variables or domains are not included. For agricultural regions, this might include the extent of irrigation; for urban areas, better measures of education could be important. The measure of unmanaged land does not account for the potential usefulness of that land.

However, comparative measures such as these can be an important first step toward determining where to direct resources—for further analysis or additional factors.

Conclusions: High Risk Impacts

Energy

Russia's current and foreseeable future economic health depends extremely heavily on Russia's ability to produce and export oil and natural gas. The oil and gas infrastructure that exists today was not designed with an eye to vulnerability stemming from a changing climate—such as structural subsidence, pipeline crossings at surging rivers, and sea-level rise. Therefore, current production will be increasingly at risk in the coming years. Moreover, as production from traditional Russian gas supply provinces declines, Russia must develop replacement sources. The Yamal Peninsula and Barents Sea (including the Shtokman field) are absolute priorities, and the Russian Far East is a secondary priority. The impacts of a changing climate may delay significantly (or significantly raise the cost of) efforts to bring these new production areas on-line, which could affect Russia's fiscal position and balance of payments.

Agriculture

By 2030, Russia will begin to experience significant changes in agriculture. The critical question will be whether the positive impacts – longer growing season, new land that can be put under the plow, and the possibility of introducing new varieties and new crops—will outweigh the significant negative impacts. In this latter regard, the reduction in precipitation in parts of Russia's traditional agricultural belts and the projected reduction in yields for traditional grain crops are significant considerations. Also significant is the projected increasing reliance on irrigation and chemical additives to deter pests and enrich soils. Rural Russia historically has been a very traditionalist part of the country, and to date Russia has not developed widespread systems to educate farmers, particularly to help them anticipate and adapt to changes in their growing conditions stemming from climate change. There are risks that rural Russia simply will not adapt itself in a timely manner to the agricultural realities of a changing climate. Russia's food supply could be under stress.

Migration

Many of Russia's southern neighbors face a drier, hotter future in which economic prospects may become increasingly dire. If these neighbors are unable to adapt themselves in a timely manner and provide for their populations, Russia may experience significant new migration pressures, which could plausibly be associated with greater instability and ethnic strife in affected Russian cities and towns.

Accentuated Socio-Economic and Socio-Political Stresses

Russia is a massive country with a pronounced continental climate, thus extreme weather is not entirely unfamiliar. Nor is hardship unfamiliar to the people of Russia. Nonetheless, the significant increase in dangerous weather events over the last two decades, and the prospect of a continuing trend in this regard, make clear that extreme weather may be the sword of Damocles hanging over Russia's future. Heat waves, wind storms, droughts, and severe flooding may result in considerable damage to infrastructure, impacts on livelihoods, and even significant loss of life. These threats, in turn, may place even greater socio-economic and socio-political stress on parts of the country where the relationship between the government and governed is already tense. Areas such as the North Caucasus have already seen political tensions and instability that are unrelated to climate change. By 2030, however, climate change could significantly exacerbate such areas of stress.

ANNEX A. ACCURACY OF REGIONAL MODELS

This is an excerpt from IPCC (2007), Chapter 11, Regional models; see IPCC 2007 for references.[4]

---

[4] Some references in this section have been changed to be internally consistent with this document and other references have been removed to avoid confusion.

11.4.2. Skill of Models in Simulating Present Climate

Regional mean temperature and precipitation in the MMD models show biases when compared with observed climate (Table 2). The multi-model mean shows a cold and wet bias in all regions and in most seasons, and the bias of the annual average temperature ranges from –2.5°C over the Tibetan Plateau (TIB) to –1.4°C over South Asia (SAS). For most regions, there is a 6°C to 7°C range in the biases from individual models with a reduced bias range in Southeast Asia (SEA) of 3.6°C. The median bias in precipitation is small (less than 10 percent) in Southeast Asia, South Asia, and Central Asia (CAS), larger in northern Asia and East Asia (NAS and EAS, around +23 percent), and very large in the Tibetan Plateau (+110 percent). Annual biases in individual models are in the range of –50 to +60 percent across all regions except the Tibetan Plateau, where some models simulate annual precipitation 2.5 times that observed and even larger seasonal biases occur in winter and spring. These global models clearly have significant problems over Tibet, due to the difficulty in simulating the effects of the dramatic topographic relief, as well as the distorted albedo feedbacks due to extensive snow cover. However, with only limited observations available, predominantly in valleys, large errors in temperature and significant underestimates of precipitation are likely.

South Asia

Over South Asia, the summer is dominated by the southwest monsoon, which spans the four months from June to September and dominates the seasonal cycles of the climatic parameters. While most models simulate the general migration of seasonal tropical rain, the observed maximum rainfall during the monsoon season along the west coast of India, the north Bay of Bengal and adjoining northeast India is poorly simulated by many models (Lal and Harasawa, 2001; Rupa Kumar and Ashrit, 2001; Rupa Kumar et al., 2002, 2003). This is likely linked to the coarse resolution of the models, as the heavy rainfall over these regions is generally associated with the steep orography. However, the simulated annual cycles in South Asian mean precipitation and surface air temperature are reasonably close to the observed. The MMD models capture the general regional features of the monsoon, such as the low rainfall amounts coupled with high variability over northwest India. However, there has not yet been sufficient analysis of whether finer details of regional significance are simulated more adequately in the MMD models.

Recent work indicates that time-slice experiments using an AGCM with prescribed SSTs, as opposed to a fully coupled system, are not able to accurately capture the South Asian monsoon response (Douville, 2005). Thus, neglecting the short-term SST feedback and variability seems to have a significant impact on the projected monsoon response to global warming, complicating the regional downscaling problem. However, May (2004a) notes that the high-resolution (about 1.5 degrees) European Centre-Hamburg (ECHAM4) GCM simulates the variability and extremes of daily rainfall (intensity as well as frequency of wet days) in good agreement with the observations (Global Precipitation Climatology Project, Huffman et al., 2001).

Three-member ensembles of baseline simulations (1961–1 990) from an RCM (PRECIS) at 50-kilometer resolution have confirmed that significant improvements in the representation of regional processes over South Asia can be achieved (Rupa Kumar et al., 2006). For example, the steep gradients in monsoon precipitation with a maximum along the western coast of India are well represented in PRECIS.

East Asia

Simulated temperatures in most MMD models are too low in all seasons over East Asia; the mean cold bias is largest in winter and smallest in summer. Zhou and Yu (2006) show that over China, the models perform reasonably in simulating the dominant variations of the mean temperature over China, but not the spatial distributions. The annual precipitation over East Asia exceeds the observed estimates in almost all models and the rain band in the mid-latitudes is shifted northward in seasons other than summer. This bias in the placement of the rains in central China also occurred in earlier models (e.g., Zhou and Li, 2002; Gao et al., 2004). In winter, the area-mean precipitation is overestimated by more than 50 percent on average due to strengthening of the rain band associated with extratropical systems over South China. The bias and inter-model differences in precipitation are smallest in summer but the northward shift of this rain band results in large discrepancies in summer rainfall distribution over Korea, Japan and adjacent seas.

Kusunoki et al. (2006) find that the simulation of the Meiyu-Changma-Baiu rains in the East Asian monsoon is improved substantially with increasing horizontal resolution. Confirming the importance of resolution, RCMs simulate more realistic climatic characteristics over East Asia than AOGCMs, whether driven by re-analyses or by AOGCMs (e.g., Ding et al., 2003; Oh et al., 2004; Fu et al., 2005; Zhang et al., 2005a, Ding et al., 2006; Sasaki et al., 2006b). Several studies reproduce the fine-scale climatology of small areas using a multiply nested RCM (Im et al., 2006) and a very-high resolution (5 kilometers) RCM (Yasunaga et al., 2006). Gao et al. (2006b) report that simulated East Asia large-scale precipitation patterns are significantly affected by resolution, particularly during the mid- to late-monsoon months, when smaller-scale convective processes dominate.

Southeast Asia

The broad-scale spatial distribution of temperature and precipitation in DJF and JJA averaged across the MMD models compares well with observations. Rajendran et al. (2004) examine the simulation of current climate in the MRI coupled model. Large-scale features were well simulated, but errors in the timing of peak rainfall over Indochina were considered a major shortcoming. Collier et al. (2004) assess the performance of the CCSM3 model in simulating tropical precipitation forced by observed SST. Simulation was good over the Maritime continent compared to the simulation for other tropical regions. B. Wang et al. (2004) assess the ability of 11 AGCMs in the Asian-Australian monsoon region simulation forced with observed SST variations. They found that the models' ability to simulate observed interannual rainfall variations was poorest in the Southeast Asian portion of the domain. Since current AOGCMs continue to have some significant shortcomings in representing ENSO variability, the difficulty of projecting changes in ENSO-related rainfall in this region is compounded.

Rainfall simulation across the region at finer scales has been examined in some studies. The Commonwealth Scientific and Industrial Research Organisation (CSIRO) stretched- grid Conformal-Cubic Atmospheric Model (CCAM) at 80-kilometer resolution shows reasonable precipitation simulation in JJA, although Indochina tended to be drier than in the observations (McGregor and Nguyen, 2003). Aldrian et al. (2004a) conducted a number of simulations with the Max-Planck Institute (MPI) regional model for an Indonesian domain, forced by reanalyses and by the ECHAM4 GCM. The model was able to represent the spatial pattern of seasonal rainfall. It was found that a resolution of at least 50 kilometers was required to simulate rainfall seasonality correctly over Sulawesi. The formulation of a coupled regional model improves regional rainfall simulation over the oceans (Aldrian et al., 2004b). Arakawa and Kitoh (2005) demonstrate an accurate simulation of the diurnal cycle of rainfall over Indonesia with an AGCM of 20-kilometer horizontal resolution.

Table 2. Biases in Present-Day (1980-1999) Surface Air Temperature and Precipitation in the MMD Simulations. The Simulated Temperatures are Compared with the HadCRUT2v (Jones, et al., 2001) Data Set and Precipitation with the CMAP (Update of Xie and Arkin, 1997) Data Set

		temperature BIAS					% precipitation BIAS				
REGION	SEASON	MIN	25	50	75	MAX	MIN	25	50	75	MAX
Asia											
NAS	DJF	-9.3	-2.9	-1.3	0.0	2.9	-18	5	12	19	93
	MAM	-6.8	-4.3	-2.7	-0.5	0.8	-4	39	45	74	110
	JJA	-4.8	-2.0	-0.5	0.4	2.2	-38	-2	19	32	62
	SON	-6.2	-2.6	-2.1	-0.5	1.9	-14	12	23	30	49
	ANN	-5.2	-2.6	-1.4	-0.6	1.3	-11	15	24	35	56
CAS	DJF	-4.4	-2.6	-1.2	0.2	3.3	-33	-2	18	43	77
	MAM	-4.3	-3.0	-1.4	0.2	2.0	-30	22	25	34	93
	JJA	-4.9	-1.6	0.3	1.4	5.7	-71	-37	-25	14	80
	SON	-4.5	-3.2	-1.9	-0.4	1.8	49	-12	-4	15	47
	ANN	-3.9	-2.3	-1.4	0.6	2.2	-44	4	12	21	53
TIB	DJF	-9.3	-3.8	-2.2	-1.4	2.2	16	131	177	255	685
	MAM	-7.3	-4.3	-3.8	-1.3	0.6	130	160	209	261	486
	JJA	-6.7	-2.5	-1.0	-0.2	1.6	4	30	37	53	148
	SON	-5.9	-3.6	-2.5	-1.7	0.0	66	93	150	180	330
	ANN	-5.3	-3.3	-2.5	-1.6	0.8	51	88	110	142	244
EAS	DJF	-6.5	-4.5	-3.7	-1.3	1.8	-20	26	60	79	142
	MAM	-5.2	-2.9	-2.0	-1.0	0.5	1	32	45	60	105
	JJA	-3.9	-2.0	-1.1	-0.4	1.4	-16	0	3	15	27
	SON	-5.9	-3.4	-2.7	-1.6	-0.3	-17	1	14	34	75
	ANN	-5.4	-3.2	-2.5	-1.2	0.2	-8	12	22	31	60
SAS	DJF	-7.4	-4.0	-2.6	-1.6	1.9	-27	0	30	59	127
	MAM	-5.6	-1.9	-0.7	-0.4	2.5	-44	-26	-1	13	72
	JJA	-2.9	-1.3	-0.1	0.6	1.9	-70	-25	-14	5	29
	SON	-5.2	-3.2	-2.1	-0.9	2.6	-26	-12	-2	14	42
	ANN	-4.8	-2.4	-1.4	-0.8	2.2	-49	-16	-10	5	33
SEA	DJF	-3.5	-2.6	-1.8	-1.2	0.4	-37	-10	-2	26	49
	MAM	-2.8	-1.6	-0.5	-0.1	1.1	-32	-9	11	25	59
	JJA	-2.5	-1.8	-0.7	-0.4	1.3	-28	-10	4	16	48
	SON	-3.3	-1.9	-1.2	-0.8	1.0	-37	-12	-4	18	51
	ANN	-2.8	-1.9	-1.0	-0.5	0.8	-28	-13	0	23	43

Temperature biases are in °C and precipitation biases in per cent. Shown are the minimum, median (50%) and maximum biases among the models, as well as the first (25%) and third (75%) quartile values. Colors indicate regions/seasons for which at least 75% of the models have the same sign of bias, with orange indicating positive and light violet negative temperature biases and light blue positive and light brown negative precipitation biases.

Central Asia and Tibet

Due to the complex topography and the associated mesoscale weather systems of the high-altitude and arid areas, GCMs typically perform poorly over the region. Importantly, the GCMs, and to a lesser extent RCMs, tend to overestimate the precipitation over arid and semi-arid areas in the north (e.g., Small et al., 1999; Gao et al., 2001; Elguindi and Giorgi, 2006).

Over Tibet, the few available RCM simulations generally exhibit improved performance in the simulation of present-day climate compared to GCMs (e.g., Gao et al., 2003a,b; Zhang et al., 2005b). For example, the GCM simulation of Gao et al. (2003a) overestimated the precipitation over the north-western Tibetan Plateau by a factor of five to six, while in an RCM nested in this model, the overestimate was less than a factor of two.

Annex B. Knowledge Deficiencies that Preclude a Full Evaluation of Climate Change Impacts on Russia and Russia's Adaptive Capacity

In order to increase the likelihood that this evaluation represents a reasonable assessment of Russia's projected climate changes and their impacts, and the country's adaptive capacity, the following gaps would need to be addressed:

- In physical science research, regional analyses will continue to be limited by the inability to model regional climates satisfactorily, including complexities arising from the interaction of global, regional, and local processes. One gap of particular interest is the lack of medium-term (20-30 years) projections that could be relied upon for planning purposes. Similarly, scientific projections of water supply and agricultural productivity are limited by inadequate understanding of various climate and physical factors affecting both areas. Research agendas in these areas can be found in, for instance, the synthesis and assessment reports of the US Climate Change Science Program (http://www.climatescience.gov) and the National Academy of Sciences (e.g., http://books.nap.edu/catalog.php?record_ id=11175 #toc). Similar types of issues exist for the biological and ecological systems that are affected.
- In social science research, scientists and analysts have only partial understandings of the important factors in vulnerability, resilience, and adaptive capacity—much less their interactions and evolution. Again, research agendas on vulnerability, adaptation, and decision-

making abound (e.g., (http://books.nap.edu/catalog.php?record_id=12545).

- Important factors are unaccounted for in research; scientists know what some of them are, but there are likely factors whose influence will be surprising. An example from earlier research on the carbon cycle illustrates this situation. The first carbon cycle models did not include carbon exchanges involving the terrestrial domain. Modelers assumed that the exchange was about equal, and the only factor modeled was deforestation. This assumption, of course, made the models inadequate for their purposes. In another example, ecosystems research models are only beginning to account for changes in pests, e.g., the pine bark beetle.
- Social models or parts of models in climate research have been developed to simulate consumption (with the assumption of well-functioning markets and rational actor behavior) and mitigation/ adaptation policies (but without attention to the social feasibility of enacting or implementing such policies). As anthropogenic climate change is the result of human decisions, the lack of knowledge about motivation, intent, and behavior is a serious lack.

Overall, research about climate change impacts on Russia has been undertaken piecemeal: discipline by discipline, sector by sector, with political implications separately considered from physical effects. This knowledge gap can be remedied by integrated research into energy-economic-environmental-political conditions and possibilities.

End Notes

[i] The Library of Congress Country Studies, http://lcweb2.loc.gov/frd/cs/ (accessed February 5, 2009); CIA World Factbook, https://www.cia.gov/library/publications/the-world-factbook/ (accessed February 5, 2009); "Russia Climate," http://www.russiansabroad.com/russian_history_96.html (accessed February 5, 2009).

[ii] J. Overpeck, K. Hughen, D. Hardy, R. Bradley, R. Case, M. Douglas, B. Finney, K. Gajewski, G. Jacoby, A. Jennings, S. Lamoueastreux, A. Lasca, G. MacDonald, J. Moore, M. Retelle, S. Smith, A. Wolfe, G. Zielinski, "Arctic Environmental Change of the Last Four Centuries." *Science* 278, no. 5341 (1997): 1251- 1256.

[iii] Federal Service for Hydrometeorology and Environmental Monitoring (Roshydromet), "Strategic Prediction for the Period of up to 2010-2015 of Climate Change Expected in Russia and its Impact on Sectors of the Russian National Economy" (Moscow, 2005), http://wmc.meteoinfo.ru/media (accessed February 17, 2009).

[iv] J. Overpeck, K. Hughen, D. Hardy, R. Bradley, R. Case, M. Douglas, B. Finney, K. Gajewski, G. Jacoby, A. Jennings, S. Lamoueastreux, A. Lasca, G. MacDonald, J. Moore, M. Retelle, S. Smith, A. Wolfe, G. Zielinski, "Arctic Environmental Change of the Last Four Centuries." *Science* 278, no. 5341 (1997): 1251-1256.

[v] R. E. Moritz, C.M. Bitz and E. J. Steig, "Dynamics of Recent Climate Change in the Arctic." *Science* 297 (2002): 1497–1502.

[vi] R. S. Bradley, H. F. Diaz, J. K. Eischeid, P. D. Jones, P. M. Kelly and C. M. Goodess, "Precipitation Fluctuations over Northern Hemisphere Land Areas Since the Mid-19th Century." *Science* 237 (1987): 171–175.

[vii] IPCC [Intergovernmental Panel on Climate Change]. *Climate Change 2007: The AR4 Synthesis Report*, eds. Rajendra K. Pachauri and Andy Reisinger (Cambridge: Cambridge University Press, 2007); IPCC [Intergovernmental Panel on Climate Change]. *Climate Change 2007: The Physical Science Basis*, eds. Susan Solomon, Dahe Qin, Martin Manning, Melinda Marquis, Kristen Averyt, Melinda M.B. Tignor, Henry LeRoy Miller, Jr., Zhenlin Chen (Cambridge: Cambridge University Press, 2007); IPCC [Intergovernmental Panel on Climate Change]. *Climate Change 2007: Impacts, Adaptation and Vulnerability*, eds. Martin Parry, Osvaldo Canziani, Jean Palutikof, Paul van der Linden, and Clair Hanson (Cambridge: Cambridge University Press, 2007); IPCC [Intergovernmental Panel on Climate Change]. *Climate Change 2007: Mitigation of Climate Change*, eds. Bert Metz, Ogunlade Davidson, Peter Bosch, Rutu Dave, Leo Meyer (Cambridge: Cambridge University Press, 2007).

[viii] Climate Change Risk Management Ltd, "Climate Change in Russia: research and impacts" (May 2008), http://www.uk-russia-ccproject.info/documents/Impacts_in_Russia_Report_2008.pdf (accessed February 17, 2009).

[ix] WWF [World Wildlife Fund], "The Impact of Climate Change on the Russian Arctic and Paths to Solving the Problem," Russia, Moscow, 2008.

[x] Federal Service for Hydrometeorology and Environmental Monitoring (Roshydromet), "Strategic Prediction for the Period of up to 2010-2015 of Climate Change Expected in Russia and its Impact on Sectors of the Russian National Economy" (Moscow, 2005), http://wmc.meteoinfo.ru/media (accessed February 17, 2009).

[xi] WWF [World Wildlife Fund], "The Impact of Climate Change on the Russian Arctic and Paths to Solving the Problem," Russia, Moscow, 2008.

[xii] WWF [World Wildlife Fund], "The Impact of Climate Change on the Russian Arctic and Paths to Solving the Problem," Russia, Moscow, 2008.

[xiii] R. E. Moritz, C.M. Bitz and E. J. Steig, "Dynamics of Recent Climate Change in the Artic." *Science* 297 (2002): 1497–1502.

[xiv] A contrary opinion is cited by a recent report from the United Nations Development Program (UNDP), which summarizes Russia's future climate as follows: "For Russia as a whole, climate change impact can be characterized as the warming accompanied by rising dryness." Source: United Nations Development Program (UNDP). Renat Perelet, Serguey Pegov and Mikhail Yulkin. *Human Development Report 2007/2008. Fighting climate change: Human solidarity in a divided world. Climate Change. Russia Country Paper* (2007).

[xv] R. W. Corell, "Challenges of climate change: an Arctic perspective." *Ambio* 35, no. 4 (2006): 148-152.

[xvi] Climate Change Risk Management Ltd, "Climate Change in Russia: research and impacts" (May 2008), http://www.uk-russia-ccproject.info/documents/Impacts_in_Russia_Report_2008.pdf (accessed February 17, 2009).

[xvii] J. Overpeck, K. Hughen, D. Hardy, R. Bradley, R. Case, M. Douglas, B. Finney, K. Gajewski, G. Jacoby, A. Jennings, S. Lamoueastreux, A. Lasca, G. MacDonald, J. Moore, M. Retelle, S. Smith, A. Wolfe, G. Zielinski, "Arctic Environmental Change of the Last Four Centuries." *Science* 278, no. 5341 (1997): 1251-1256.

xviii Federal Service for Hydrometeorology and Environmental Monitoring (Roshydromet), "Strategic Prediction for the Period of up to 2010-2015 of Climate Change Expected in Russia and its Impact on Sectors of the Russian National Economy" (Moscow, 2005), http://wmc.meteoinfo.ru/media (accessed February 17, 2009).

xix Source: R. E. Moritz, C.M. Bitz and E. J. Steig, *Science* 297 (2002): 1497–1502.

xx J. Overpeck, K. Hughen, D. Hardy, R. Bradley, R. Case, M. Douglas, B. Finney, K. Gajewski, G. Jacoby, A. Jennings, S. Lamoueastreux, A. Lasca, G. MacDonald, J. Moore, M. Retelle, S. Smith, A. Wolfe, G. Zielinski, "Arctic Environmental Change of the Last Four Centuries." *Science* 278, no. 5341 (1997): 1251-1256.

xxi Climate Change Risk Management Ltd, "Climate Change in Russia: research and impacts" (May 2008), http://www.uk-russia-ccproject.info/documents/Impacts_in_Russia_Report_2008.pdf (accessed February 17, 2009).

xxii V. P. Meleshko, et al. "Climate of Russia in the 21st Century: Part 3. Future Climate Changes Calculated with an Ensemble of Coupled Atmosphere–Ocean General Circulation CMIP3 Models." *Russian Meteorology and Hydrology*

xxiii V. P. Meleshko, et al. "Climate of Russia in the 21st Century: Part 3. Future Climate Changes Calculated with an Ensemble of Coupled Atmosphere–Ocean General Circulation CMIP3 Models." *Russian Meteorology and Hydrology* 33, no. 9 (2008): 541–552, ISSN: 1068-3739.

xxiv WWF [World Wildlife Fund], "The Impact of Climate Change on the Russian Arctic and Paths to Solving the Problem," Russia, Moscow, 2008.

xxv WWF [World Wildlife Fund], "The Impact of Climate Change on the Russian Arctic and Paths to Solving the Problem," Russia, Moscow, 2008.

xxvi J. Overpeck, K. Hughen, D. Hardy, R. Bradley, R. Case, M. Douglas, B. Finney, K. Gajewski, G. Jacoby, A. Jennings, S. Lamoueastreux, A. Lasca, G. MacDonald, J. Moore, M. Retelle, S. Smith, A. Wolfe, G. Zielinski, "Arctic Environmental Change of the Last Four Centuries." *Science* 278, no. 5341 (1997): 1251-1256.

xxvii J. Overpeck, K. Hughen, D. Hardy, R. Bradley, R. Case, M. Douglas, B. Finney, K. Gajewski, G. Jacoby, A. Jennings, S. Lamoueastreux, A. Lasca, G. MacDonald, J. Moore, M. Retelle, S. Smith, A. Wolfe, G. Zielinski, "Arctic Environmental Change of the Last Four Centuries." *Science* 278, no. 5341 (1997): 1251-1256.

xxviii Climate Change Risk Management Ltd, "Climate Change in Russia: research and impacts" (May 2008), http://www.uk-russia-ccproject.info/documents/Impacts_in_Russia_Report_2008.pdf (accessed March 31, 2009).

xxix Climate Change Risk Management Ltd, "Climate Change in Russia: research and impacts" (May 2008), http://www.uk-russia-ccproject.info/documents/Impacts_in_Russia_Repor_2008.pdf (accessed March 31, 2009).

xxx R. E. Moritz, C.M. Bitz and E. J. Steig, "Dynamics of Recent Climate Change in the Artic." *Science* 297 (2002): 1497–1502.

xxxi Climate Change Risk Management Ltd, "Climate Change in Russia: research and impacts" (May 2008), http://www.uk-russia-ccproject.info/documents/Impacts_in_Russia_Report_2008.pdf (accessed March 31, 2009).

xxxii Oleg Anisimov and Svetlana Reneva. "Permafrost and Changing Climate: The Russian Perspective." *AMBIO: A Journal of the Human Environment* 35, no. 4 (2006): 169–175.

xxxiii Federal Service for Hydrometeorology and Environmental Monitoring (Roshydromet), "Strategic Prediction for the Period of up to 2010-2015 of Climate Change Expected in Russia and its Impact on Sectors of the Russian National Economy" (Moscow, 2005), http://wmc.meteoinfo.ru/media (accessed February 17, 2009).

xxxiv J. Overpeck, K. Hughen, D. Hardy, R. Bradley, R. Case, M. Douglas, B. Finney, K. Gajewski, G. Jacoby, A. Jennings, S. Lamoueastreux, A. Lasca, G. MacDonald, J. Moore, M. Retelle, S. Smith, A. Wolfe, G. Zielinski, "Arctic Environmental Change of the Last Four Centuries." *Science* 278, no. 5341 (1997): 1251-1256.

[xxxv] R. S. Bradley, H. F. Diaz, J. K. Eischeid, P. D. Jones, P. M. Kelly and C. M. Goodess, "Precipitation Fluctuations over Northern Hemisphere Land Areas since the Mid-19th Century." *Science* 237 (1987): 171–175.

[xxxvi] Federal Service for Hydrometeorology and Environmental Monitoring (Roshydromet), "Strategic Prediction for the Period of up to 2010-2015 of Climate Change Expected in Russia and its Impact on Sectors of the Russian National Economy" (Moscow, 2005), http://wmc.meteoinfo.ru/media (accessed February 17, 2009).

[xxxvii] One presumes that "climate-related" yield means that other variables have been factored out; however, this is not explicitly stated in the report.

[xxxviii] Federal Service for Hydrometeorology and Environmental Monitoring (Roshydromet), "Strategic Prediction for the Period of up to 2010-2015 of Climate Change Expected in Russia and its Impact on Sectors of the Russian National Economy" (Moscow, 2005), http://wmc.meteoinfo.ru/media (accessed February 17, 2009).

[xxxix] Federal Service for Hydrometeorology and Environmental Monitoring (Roshydromet), "Strategic Prediction for the Period of up to 2010-2015 of Climate Change Expected in Russia and its Impact on Sectors of the Russian National Economy" (Moscow, 2005), http://wmc.meteoinfo.ru/media (accessed February 17, 2009).

[xl] IPCC [Intergovernmental Panel on Climate Change]. *Climate Change 2007: Impacts, Adaptation and Vulnerability*, eds. Martin Parry, Osvaldo Canziani, Jean Palutikof, Paul van der Linden, and Clair Hanson (Cambridge: Cambridge University Press, 2007)

[xli] IPCC [Intergovernmental Panel on Climate Change]. *Climate Change 2007: Impacts, Adaptation and Vulnerability*, eds. Martin Parry, Osvaldo Canziani, Jean Palutikof, Paul van der Linden, and Clair Hanson (Cambridge: Cambridge University Press, 2007)

[xlii] Eric F. Lambin and Helmut J. Geist, eds., *Land-Use And Land-Cover Change: Local Processes And Global Impacts* (Berlin, Heidelberg: Springer, 2006).

[xliii] R. S. Bradley, H. F. Diaz, J. K. Eischeid, P. D. Jones, P. M. Kelly and C. M. Goodess, Precipitation Fluctuations over Northern Hemisphere Land Areas Since the Mid-19th Century." *Science* 237 (1987): 171–175.

[xliv] Federal Service for Hydrometeorology and Environmental Monitoring (Roshydromet), "Strategic Prediction for the Period of up to 2010-2015 of Climate Change Expected in Russia and its Impact on Sectors of the Russian National Economy" (Moscow, 2005), http://wmc.meteoinfo.ru/media (accessed February 17, 2009).

[xlv] Renat Perelet, Serguey Pegov and Mikhail Yulkin. *Human Development Report 2007/2008. Fighting climate change: Human solidarity in a divided world. Climate Change. Russia Country Paper* (2007), United Nations Development Program (UNDP).

[xlvi] J. Overpeck, K. Hughen, D. Hardy, R. Bradley, R. Case, M. Douglas, B. Finney, K. Gajewski, G. Jacoby, A. Jennings, S. Lamoueastreux, A. Lasca, G. MacDonald, J. Moore, M. Retelle, S. Smith, A. Wolfe, G. Zielinski, "Arctic Environmental Change of the Last Four Centuries." *Science* 278, no. 5341 (1997): 1251-1256.

[xlvii] J. Overpeck, K. Hughen, D. Hardy, R. Bradley, R. Case, M. Douglas, B. Finney, K. Gajewski, G. Jacoby, A. Jennings, S. Lamoueastreux, A. Lasca, G. MacDonald, J. Moore, M. Retelle, S. Smith, A. Wolfe, G. Zielinski, "Arctic Environmental Change of the Last Four Centuries." *Science* 278, no. 5341 (1997): 1251-1256.

[xlviii] Climate Change Risk Management Ltd. May 2008. "Climate Change in Russia: research and impacts." http://www.uk-russia-ccproject.info/documents/Impacts_in_Russia_Report_2008.pdf (accessed February 17, 2009).

[xlix] Climate Change Risk Management Ltd. May 2008. "Climate Change in Russia: research and impacts." http://www.uk-russia-ccproject.info/documents/Impacts_in_Russia_Report_2008.pdf (accessed February 17, 2009).

[l] Climate Change Risk Management Ltd. May 2008. "Climate Change in Russia: research and impacts." http://www.uk-russia-ccproject.info/documents/Impacts_in_Russia_Report_2008.pdf (accessed February 17, 2009).

[li] Climate Change Risk Management Ltd. May 2008. "Climate Change in Russia: research and impacts." http://www.uk-russia-ccproject.info/documents/Impacts_in_Russia_Report_2008.pdf (accessed February 17, 2009).

[lii] R. S. Bradley, H. F. Diaz, J. K. Eischeid, P. D. Jones, P. M. Kelly and C. M. Goodess, Precipitation Fluctuations over Northern Hemisphere Land Areas Since the Mid-19th Century." *Science* 237 (1987): 171–175.

[liii] IPCC [Intergovernmental Panel on Climate Change]. *Climate Change 2007: The AR4 Synthesis Report*, eds. Rajendra K. Pachauri and Andy Reisinger (Cambridge: Cambridge University Press, 2007)

[liv] Federal Service for Hydrometeorology and Environmental Monitoring (Roshydromet), "Strategic Prediction for the Period of up to 2010-2015 of Climate Change Expected in Russia and its Impact on Sectors of the Russian National Economy" (Moscow, 2005), http://wmc.meteoinfo.ru/media (accessed February 17, 2009).

[lv] Climate Change Risk Management Ltd, "Climate Change in Russia: research and impacts" (May 2008), http://www.uk-russia-ccproject.info/documents/Impacts_in_Russia_Report_2008.pdf (accessed February 17, 2009).

[lvi] R. E. Moritz, C.M. Bitz and E. J. Steig, "Dynamics of Recent Climate Change in the Artic." *Science* 297 (2002): 1497–1502.

[lvii] R. E. Moritz, C.M. Bitz and E. J. Steig, "Dynamics of Recent Climate Change in the Artic." *Science* 297 (2002): 1497–1502.

[lviii] Climate Change Risk Management Ltd, "Climate Change in Russia: research and impacts" (May 2008), http://www.uk-russia-ccproject.info/documents/Impacts_in_Russia_Report_2008.pdf (accessed February 17, 2009).

[lix] Federal Service for Hydrometeorology and Environmental Monitoring (Roshydromet), "Strategic Prediction for the Period of up to 2010-2015 of Climate Change Expected in Russia and its Impact on Sectors of the Russian National Economy" (Moscow, 2005), http://wmc.meteoinfo.ru/media (accessed February 17, 2009).

[lx] Federal Service for Hydrometeorology and Environmental Monitoring (Roshydromet), "Strategic Prediction for the Period of up to 2010-2015 of Climate Change Expected in Russia and its Impact on Sectors of the Russian National Economy" (Moscow, 2005), http://wmc.meteoinfo.ru/media (accessed February 17, 2009); Also: Federal Service for Hydrometeorology and Environmental Monitoring (Roshydromet), "Assessment Report on Climate Change and its Consequences in Russian Federation" (Moscow, 2008).

[lxi] Federal Service for Hydrometeorology and Environmental Monitoring (Roshydromet), "Assessment Report on Climate Change and its Consequences in Russian Federation" (Moscow, 2008).

[lxii] Many Russian cities employ district heating systems that are turned on at the beginning of the season and operate at constant output levels until the end of the heating season. Consequently, there is limited scope to respond to a brief warm spell in the middle of the heating season by producing less heat. For the discussion of unrealized fuel savings, see: Federal Service for Hydrometeorology and Environmental Monitoring (Roshydromet), "Strategic Prediction for the Period of up to 2010-2015 of Climate Change Expected in Russia and its Impact on Sectors of the Russian National Economy" (Moscow, 2005), http://wmc.meteoinfo.ru/media (accessed February 17, 2009).

[lxiii] Government of the Russian Federation, "Chetvertoye Natsional'noye Soobshcheniye predstavlyaemoye v sootvetstviyi so stat'yami 4 i 12 Ramochnoy Konventsii Organizatsii Ob'yedinyennikh Natsii ob izmenenii klimata i stat'yey 7 Kyotskogo protokola" ("Fourth National Communication presented in accordance with articles 4 and 12 of the United Nations Framework Convention on Climate Change and article 7 of the Kyoto Protocol" – in Russian) (Moscow: 2006).

[lxiv] Federal Service for Hydrometeorology and Environmental Monitoring (Roshydromet), "Strategic Prediction for the Period of up to 2010-2015 of Climate Change Expected in Russia and its Impact on Sectors of the Russian National Economy" (Moscow, 2005),

http://wmc.meteoinfo.ru/media (accessed February 17, 2009).

[lxv] Federal Service for Hydrometeorology and Environmental Monitoring (Roshydromet), "Strategic Prediction for the Period of up to 2010-2015 of Climate Change Expected in Russia and its Impact on Sectors of the Russian National Economy" (Moscow, 2005), http://wmc.meteoinfo.ru/media (accessed February 17, 2009).

[lxvi] Clifford Gaddy, Fiona Hill, Igor Danchenko, "The Russian Federation – The Brookings Foreign Policy Studies Energy Security Series," (Washington, DC: Brookings Institution, October 2006).

[lxvii] Clifford Gaddy and Barry Ickes. "Putin's Third Way," (The National Interest Online, January 21, 2009), http://www.nationalinterest.org/Article.aspx?id=20496.

[lxviii] Federal Service for Hydrometeorology and Environmental Monitoring (Roshydromet), "Strategic Prediction for the Period of up to 2010-2015 of Climate Change Expected in Russia and its Impact on Sectors of the Russian National Economy" (Moscow, 2005), http://wmc.meteoinfo.ru/media (accessed February 17, 2009).

[lxix] Federal Service for Hydrometeorology and Environmental Monitoring (Roshydromet), "Strategic Prediction for the Period of up to 2010-2015 of Climate Change Expected in Russia and its Impact on Sectors of the Russian National Economy" (Moscow, 2005), http://wmc.meteoinfo.ru/media (accessed February 17, 2009).

[lxx] International Energy Agency, *Optimising Russian Natural Gas: Reform and Climate Policy* (Paris: OECD/IEA, 2006).

[lxxi] Already before the financial crisis hit in late 2008, it was clear that to industry analysts that new Gazprom production is lagging behind the company's ambitious targets. Source: International Energy Agency (IEA 2006), *Optimising Russian Natural Gas: Reform and Climate Policy* (Paris: OECD/IEA, 2006).

[lxxii] Jonathan Stern cites more modest and realistic projections for Yamal production. See: Jonathan P. Stern, *The Future of Russian Gas and Gazprom*, (Oxford: Oxford University Press, 2005). Gazprom itself is even more optimistic than Stern is – and publicly cites the goal of 360 bcm/year production by 2030. See: Gazprom press release, "Gazprom launches Yamal Megaproject," (December 3, 2008), http://www.gazprom.ru/eng/news/2008/12/32740.shtml.

[lxxiii] Jonathan P. Stern, *The Future of Russian Gas and Gazprom*, (Oxford: Oxford University Press, 2005).

[lxxiv] Ministry of Industry and Energy of the Russian Federation and Institute for Energy Strategy, "Kontseptsiya Energeticheskoy strategii Rossii do 2030 g. (proyekt)" ("Concept for the Energy Strategy of Russia to 2030 (draft)"—in Russian), (Moscow: 2007).

[lxxv] Federal Service for Hydrometeorology and Environmental Monitoring (Roshydromet), "Strategic Prediction for the Period of up to 2010-2015 of Climate Change Expected in Russia and its Impact on Sectors of the Russian National Economy" (Moscow, 2005), http://wmc.meteoinfo.ru/media (accessed February 17, 2009).

[lxxvi] Federal Service for Hydrometeorology and Environmental Monitoring (Roshydromet), "Assessment Report on Climate Change and its Consequences in Russian Federation" (Moscow, 2008).

[lxxvii] Federal Service for Hydrometeorology and Environmental Monitoring (Roshydromet), "Assessment Report on Climate Change and its Consequences in Russian Federation" (Moscow, 2008).

[lxxviii] Federal Service for Hydrometeorology and Environmental Monitoring (Roshydromet), "Strategic Prediction for the Period of up to 2010-2015 of Climate Change Expected in Russia and its Impact on Sectors of the Russian National Economy" (Moscow, 2005), http://wmc.meteoinfo.ru/media (accessed February 17, 2009).

[lxxix] Federal Service for Hydrometeorology and Environmental Monitoring (Roshydromet), "Strategic Prediction for the Period of up to 2010-2015 of Climate Change Expected in Russia and its Impact on Sectors of the Russian National Economy" (Moscow, 2005), http://wmc.meteoinfo.ru/media (accessed February 17, 2009).

[lxxx] Federal Service for Hydrometeorology and Environmental Monitoring (Roshydromet), "Strategic Prediction for the Period of up to 2010-2015 of Climate Change Expected in Russia and its Impact on Sectors of the Russian National Economy" (Moscow, 2005), http://wmc.meteoinfo.ru/media (accessed February 17, 2009).

[lxxxi] Federal Service for Hydrometeorology and Environmental Monitoring (Roshydromet), "Assessment Report on Climate Change and its Consequences in Russian Federation" (Moscow, 2008).

[lxxxii] Igor E. Chestin and Nicholas A. Colloff, eds, "Russia and Neighboring Countries: Environmental, Economic, and Social Impacts." (Moscow: World Wildlife Fund-Russia (WWF-Russia) and Oxfam GB, 2008). See also: Federal Service for Hydrometeorology and Environmental Monitoring (Roshydromet), "Strategic Prediction for the Period of up to 2010-2015 of Climate Change Expected in Russia and its Impact on Sectors of the Russian National Economy" (Moscow, 2005), http://wmc.meteoinfo.ru/media (accessed February 17, 2009).

[lxxxiii] Federal Service for Hydrometeorology and Environmental Monitoring (Roshydromet), "Strategic Prediction for the Period of up to 2010-2015 of Climate Change Expected in Russia and its Impact on Sectors of the Russian National Economy" (Moscow, 2005), http://wmc.meteoinfo.ru/media (accessed February 17, 2009).

[lxxxiv] Federal Service for Hydrometeorology and Environmental Monitoring (Roshydromet), "Strategic Prediction for the Period of up to 2010-2015 of Climate Change Expected in Russia and its Impact on Sectors of the Russian National Economy" (Moscow, 2005), http://wmc.meteoinfo.ru/media (accessed February 17, 2009).

[lxxxv] Federal Service for Hydrometeorology and Environmental Monitoring (Roshydromet), "Assessment Report on Climate Change and its Consequences in Russian Federation" (Moscow, 2008).

[lxxxvi] Federal Service for Hydrometeorology and Environmental Monitoring (Roshydromet), "Assessment Report on Climate Change and its Consequences in Russian Federation" (Moscow, 2008).

[lxxxvii] Nikolai Dronin and Andrei Kirilenko, "Climate change and food stress in Russia: what if the market transforms as it did during the past century?" *Climatic Change* 86 (2008): 123-150.

[lxxxviii] Federal Service for Hydrometeorology and Environmental Monitoring (Roshydromet), "Strategic Prediction for the Period of up to 2010-2015 of Climate Change Expected in Russia and its Impact on Sectors of the Russian National Economy" (Moscow, 2005), http://wmc.meteoinfo.ru/media (accessed February 17, 2009).

[lxxxix] Central Intelligence Agency: World Factbook 2008, https://www.cia.gov/library/publications/theworld-factbook/index.html.

[xc] Russian Railways company website, http://www.eng.rzd.ru/wps/portal/rzdeng?STRUCTURE_ID=4.

[xci] Roshydromet says the problem of reduced availability of ice roads will particularly affect the Russian Far East, Magadan, Sakha (Yakutia), and Chukotka. See: Federal Service for Hydrometeorology and Environmental Monitoring (Roshydromet), "Strategic Prediction for the Period of up to 2010-2015 of Climate Change Expected in Russia and its Impact on Sectors of the Russian National Economy" (Moscow, 2005), http://wmc.meteoinfo.ru/media (accessed February 17, 2009).

[xcii] Federal Service for Hydrometeorology and Environmental Monitoring (Roshydromet), "Strategic Prediction for the Period of up to 2010-2015 of Climate Change Expected in Russia and its Impact on Sectors of the Russian National Economy" (Moscow, 2005), http://wmc.meteoinfo.ru/media (accessed February 17, 2009).

[xciii] Arctic Climate Impact Assessment (ACIA), *Impacts of a Warming Climate* (Cambridge: Cambridge University Press, 2004).

[xciv] Arctic Climate Impact Assessment (ACIA), *Impacts of a Warming Climate* (Cambridge: Cambridge University Press, 2004).

[xcv] Gerald A. Meehl, Thomas F. Stocker, et al. "Chapter 10: Global Climate Projections." In IPCC [Intergovernmental Panel on Climate Change]. *Climate Change 2007: The AR4 Synthesis Report*, eds. Rajendra K. Pachauri and Andy Reisinger (Cambridge: Cambridge University Press, 2007); Federal Service for Hydrometeorology and Environmental Monitoring (Roshydromet), "Assessment Report on Climate Change and its Consequences in Russian Federation" (Moscow, 2008).

[xcvi] Cheryl Lyn Dybas, "Russia Sees Ill Effects of 'General Winter's' Retreat," *Washington Post*, April 2, 2007, A6.

[xcvii] Renat Perelet, Serguey Pegov and Mikhail Yulkin, "Climate Change. Russia Country Paper" in United Nations Development Program (UNDP), Human Development Report 2007/2008, *Fighting climate change: Human solidarity in a divided world*, 2007.

[xcviii] Federal Service for Hydrometeorology and Environmental Monitoring (Roshydromet), "Assessment Report on Climate Change and its Consequences in Russian Federation" (Moscow, 2008).

[xcix] Renat Perelet, Serguey Pegov and Mikhail Yulkin, "Climate Change. Russia Country Paper" in United Nations Development Program (UNDP), Human Development Report 2007/2008, *Fighting climate change: Human solidarity in a divided world*, 2007.

[c] Timothy L. Thomas, "EMERCOM: Russia's Emergency Response Team," *Low Intensity Conflict and Law Enforcement* 4, no. 2 (Autumn 1995): 227-236, http://www.fas.org/nuke/guide/russia/agency/rusert.htm (accessed February 16, 2009).

[ci] Federal Service for Hydrometeorology and Environmental Monitoring (Roshydromet), "Assessment Report on Climate Change and its Consequences in Russian Federation" (Moscow, 2008).

[cii] Fiona Hill and Clifford Gaddy, *The Siberian Curse* (Washington, D.C.: Brookings Institution Press, 2003).

[ciii] Federal Service for Hydrometeorology and Environmental Monitoring (Roshydromet), "Assessment Report on Climate Change and its Consequences in Russian Federation" (Moscow, 2008).

[civ] Federal Service for Hydrometeorology and Environmental Monitoring (Roshydromet), "Assessment Report on Climate Change and its Consequences in Russian Federation" (Moscow, 2008).

[cv] Igor E. Chestin and Nicholas A. Colloff, eds, "Russia and Neighboring Countries: Environmental, Economic, and Social Impacts." (Moscow: World Wildlife Fund-Russia (WWF-Russia) and Oxfam GB, 2008).

[cvi] Arctic Climate Impact Assessment (ACIA), *Impacts of a Warming Climate* (Cambridge: Cambridge University Press, 2004).

[cvii] Scott G. Borgerson, "Arctic Meltdown: The Economic and Security Implications of Global Warming," *Foreign Affairs* 87, no. 2 (March/April 2008): 63-77.

[cviii] Vladimir Isachenkov, "Russia sending more ships, scientists to Arctic," (Associated Press, February 12, 2009), http://hosted.ap.org/dynamic/stories/E/EU SECTION= HOME&TEMPLATE=DEFAULT (accessed February 16, 2009).

[cix] David Stringer, "Arctic's thaw brings security risks for NATO." (Associated Press, January 30, 2009), http://www.canadianbusiness.com/markets/headline_news/article.jsp?content=D9610E5O1 (accessed January 30, 2009).

[cx] Owen Matthews and Anna Nemtsova, "The Kremlin Vigilantes – As anti-immigrant groups grow more violent, they get more explicit support from Russian Authorities," (Newsweek International, February 23, 2009), http://www.newsweek.com/id/184777 (accessed February 20, 2009).

[cxi] Igor E. Chestin and Nicholas A. Colloff, eds., "Russia and Neighboring Countries: Environmental, Economic, and Social Impacts." (Moscow: World Wildlife Fund-Russia (WWF-Russia) and Oxfam GB, 2008).

[cxii] Anya Ardayeva, "Attacks Against Minorities Up in Russia," (Voice of America, February 19, 2009) http://www.voanews.com/english/2009-02-19-voa59.cfm (accessed February 20,

2009); Owen Matthews and Anna Nemtsova, "The Kremlin Vigilantes – As anti-immigrant groups grow more violent, they get more explicit support from Russian Authorities," (Newsweek International, February 23, 2009), http://www.newsweek.com/id/184777 (accessed February 20, 2009).

[cxiii] Intergovernmental Panel on Climate Change, "Glossary," *Climate Change 2007: Impacts, Adaptation and Vulnerability: Contribution of Working Group II to the Fourth Assessment Report of the Intergovernmental Panel on Climate Change*, eds. M.L. Parry, O.F. Canziani, J.P. Palutikof, P.J. van der Linden and C.E. Hanson (Cambridge: Cambridge University Press 2007).

[cxiv] G. Yohe and R. Tol, "Indicators for social and economic coping capacity—moving toward a working definition of adaptive capacity," *Global Environmental Change* 12 (2002): 25-40

[cxv] E.L. Malone and A.L. Brenkert, "Vulnerability, sensitivity, and coping/adaptive capacity worldwide," *The Distributional Effects of Climate Change: Social and Economic Implications*, M. Ruth and M. Ibarraran, eds., Elsevier Science, Dordrecht (in press).

[cxvi] N. Nakicenovic and R. Swart, *Special Report on Emissions Scenarios* (Cambridge: Cambridge University Press 2000).

[cxvii] Nikolai Dronin and Andrei Kirilenko, "Climate change and food stress in Russia: what if the market transforms as it did during the past century?" *Climatic Change* 86 (2008): 123-150.

In: Implications of Climate Change ... ISBN: 978-1-61122-851-9
Editor: Michelle D. Monahan et al.

Chapter 2

RUSSIA: THE IMPACT OF CLIMATE CHANGE TO 2030 - GEOPOLITICAL IMPLICATIONS*

National Intelligence Council

SCOPE NOTE

Following the publication in 2008 of the National Intelligence Assessment on the National Security Implications of Global Climate Change to 2030, the National Intelligence Council (NIC) embarked on a research effort to explore in greater detail the national security implications of climate change in six countries/regions of the world: India, China, Russia, North Africa, Mexico and the Caribbean, and Southeast Asia and the Pacific Island States. For each country/region we have adopted a three-phase approach.

- In the first phase, contracted research explores the latest scientific findings on the impact of climate change in the specific region/country. For Russia, the Phase I effort was published as a NIC Special Report: ***Russia: Impact of Climate Change to 2030, A Commissioned Research Report*** (NIC 2009-04, April 2009).
- In the second phase, a workshop or conference composed of experts from outside the Intelligence Community (IC) determines if

* This is an edited, reformatted and augmented edition of a National Intelligence Council Conference Report, Order Code CR 2009-16, dated September 2009.

anticipated changes from the effects of climate change will force inter- and intra-state migrations, cause economic hardship, or result in increased social tensions or state instability within the country/region. This chapter is the result of the Phase II effort for Russia.

- In the final phase, the NIC Long-Range Analysis Unit (LRAU) will lead an IC effort to identify and summarize for the policy community the anticipate impact on US national security.

In April of 2009, a group of regional experts convened to explore the sociopolitical challenges, civil and key interest group responses, government responses, and regional and geopolitical implications of climate change on Russia through 2030. The group of outside experts consisted of social scientists, economists, and political scientists. While the targeted time frame of the analysis was out to 2030, the perceptions of decisionmakers in 2030 will be colored by expectations about the relative severity of climate changes projected later in the century. The participants accordingly considered climate impacts beyond 2030 where appropriate.

This work is being delivered under the Global Climate Change Research Program contract with the CIA's Office of the Chief Scientists.

EXECUTIVE SUMMARY

The National Intelligence Council-sponsored workshop entitled *The Implications of Global Climate Change in Russia*, held on May 18, 2009, brought together a panel of Russia experts to consider the probable effects of climate change on Russia from a social, political, and economic perspective. **The panelists concluded that Russia is reaching a point where serious deterioration of its physical and human capital is a major obstacle to sustainable economic growth and Russia's capacity to adapt and protect its people will be tested out to 2030. Climate change will add additional stress to energy and transportation infrastructure burdens.** However, given Russia's high overall state capacity and the mixed or comparatively tolerable nature of most anticipated climate change impacts over the next twenty years, climate change is unlikely to lead to a general failure of the Russian state.

Energy is the most important determinant of Russia's economic future and state capacity and climate change will have significant direct and indirect impacts on the energy sector.

- Russia's economy is vulnerable to the uncertain effects of climate change and international climate change mitigation policies that may reduce world oil and gas prices.
- Russia will need to make massive investments in its oil and gas infrastructure, including upgrades to existing infrastructure and development of new resources. Climate change will add to the cost and technical difficulty of these projects.
- Climate change-induced effects such as permafrost melting will pose a serious threat to Russia's pipelines and other aging energy and transportation infrastructure, which is already in need of replacement.

Russia has attributes that provide it with a greater capacity to respond to the negative effects of climate change than some industrialized countries and most underdeveloped ones. The state has robust capacities in areas such as analysis and forecasting, and emergency response. Russia retains, even if aging, significant Soviet-era industrial infrastructure. The population is accustomed to privation and has developed informal networks to address shortfalls in government-provided services and resources. Demand for oil and gas in Asia will keep Russia's energy revenues, and therefore state resources, at high levels through 2030.

Russia nevertheless faces limitations that may inhibit an effective response to climate change.

- Centralization of power has diminished the capacities of local and regional governments, the normal first responders to climate change-induced challenges.
- The state prioritizes development and security with little regard for environmental issues, and a significant proportion of the leadership voices the view that a warming climate is a net benefit for Russia. Energy infrastructure demands will divert resources away from climate change adaptation and mitigation.
- Civil society—the ability for citizens to mobilize through NGOs and interest groups—is underdeveloped and civil responses to climate change, such as ecological movements, face state repression.

The considerable variation in the negative effects of climate change across the vast Russian territory also will challenge the central government's ability to respond. While climate modeling cannot reliably identify the specific effects of climate change at a local level, existing data provides insights into climatic challenges that may become most prominent in Russia's different regions.

- Melting of the Arctic permafrost threatens to undermine urban, industrial, and transportation infrastructure across northern Russia and its primary agricultural regions may face severe stress from water shortages, temperature increases, and shifts in crop zones.
- Russia will face major internal and cross-border migration, mainly into its cities, which may increase ethnic tensions. Conversely, the expected decline in the working-age population may reduce competition for employment and resources.

Russia will also face climate change-induced challenges originating outside its borders as climate change will increase competition for resources and population pressures along Russia's periphery, particularly in Central and East Asia and the Caucasus. Climatic pressures may drive migration into Russia, and the influx of migrants could stimulate xenophobia and ethnic violence.

Despite the many challenges, climate change will generate a number of benefits in some parts of Russia.

- Russia will have more water overall, which benefits hydroelectric potential, irrigation, and urban water supply.
- Warming will reduce energy demand for heating, and allow agriculture and settlement in currently inhospitable northern areas of Russia.
- Melting of the Arctic ice pack will open shipping routes along Russia's northern coast and allow access to previously inaccessible Arctic energy resources.

At Copenhagen, Moscow's probable overarching strategy will be to leave the contentious negotiations to the United States, China, India, and the European Union. Moscow will look to take full advantage of the United States' interest in brokering deals with India and China, and will take every opportunity to extract favorable concessions. Moscow may try to position itself as an important swing player and broker between the West and the developing world.

BACKGROUND AND INTRODUCTION[1]

The Russian Federation is already experiencing the impacts of climate change in the form of milder winters, melting permafrost, changing precipitation patterns, the spread of disease, and increased incidence of drought, flooding, and other extreme weather events. Many of these observed climate impacts are having concrete, negative effects on Russians' quality of life. By 2030, Russia will start to feel the impacts of climate change in relation to both water and food supply. Nonetheless, a significant proportion of the country's senior leaders continue to voice the view that a warming climate is a net benefit for Russia. Although Russia has a number of attributes that provide a greater capacity for resilience[2] in response to the negative effects of climate change than some other industrialized countries and most developing countries, as the impacts of the change continue and intensify over the coming years, Russia's capacity to adapt and protect its people will be severely tested.

The most important impacts of climate change in Russia will likely include the following:

- ***Energy.*** A warming climate holds the possibility of milder and shorter heating seasons, which in turn may lead to reduced energy demand. Increased water availability—particularly along those Siberian rivers that are used for hydroelectric power—should result in increased power production in certain parts of the country. On the other hand, existing and future energy infrastructure for the all-important petroleum industry will experience more pronounced challenges, including structural subsidence; risks associated with river crossings; and construction difficulties as permafrost thaws earlier and deeper, impeding the construction of vital new production areas. These challenges have the potential for a material, negative impact on the single greatest source of revenue to the Russian state—the oil and gas industry.

- ***Water.*** Many parts of Russia's immense territory will experience increases in the availability of water, including much of Siberia, the Far North, and northwestern Russia. This change will bring certain positive impacts, including for hydroelectricity generation. Nevertheless, managing the increased flows will pose other problems, especially when these increases coincide with extreme weather events such as downpours or springtime, ice-clogged floods. In addition, increasing water shortages are predicted for southern parts of European Russia, areas that already experience significant socioeconomic and sociopolitical stress. Moreover, a number of densely populated Russian regions that are already subject to water shortages are expected to face even more pronounced difficulties in decades to come.
- ***Agriculture.*** As growing seasons become longer and precipitation patterns change, it will become possible to cultivate northern lands that previously were too cold for agricultural purposes. It may also become possible to raise new crops and new crop result in increased yields of new or existing crops remains to be seen. Moreover, agriculture will become more reliant on irrigation (especially in the southern parts of European Russia), pesticides and herbicides, and more vulnerable to droughts and other extreme weather.
- ***Migration.*** Russia, which is already the top choice for immigrants after the United States, is likely to experience greater migration pressure from Central Asia, the Caucasus countries, Mongolia, and northeastern China.[3] The latter areas are expected to experience increased water shortages and resulting economic stress. In addition, internal migration pressures may occur as residents of Russia's many northern cities face increasing economic and climate-related challenges.
- ***Accentuation of existing socioeconomic and sociopolitical stresses.*** Russia is better equipped to deal with the impacts of climate change than many of its neighbors. Nonetheless, by 2030, climate change appears likely to accentuate some of the stresses that currently plague Russia. Some of the most affected regions already feature attenuated and unsettled socioeconomic and sociopolitical situations and most of the impacts of climate change will manifest themselves in smaller cities and in the Russian countryside. For example, the long-turbulent North Caucasus region will be drier, hotter, and less prosperous than it is today. Primorskiy Kray and the Russian Far East, which have long

struggled to develop peacefully next to China, appear likely to experience even greater migration pressures, which could exacerbate longstanding cross-border tensions.

SOCIAL, POLITICAL, AND ECONOMIC CHALLENGES

Regional Challenges

Predicting the economic, social, and political impacts of climate change across the vast territory of Russia is difficult because of the lack of micro-level climate change modeling. Affects that may appear moderate when averaged out over such a large area may in fact produce extremely intense problems at the local level that may exert disproportionate destabilizing influence. Existing data nonetheless can at least be used to identify those climatic challenges that may potentially become most prominent in Russia's different regions.

Northern European Russia

The principal climate change-induced challenges facing Russia's northwest appear to be permafrost melting, which threatens to undermine urban, industrial, and transportation infrastructure, and flooding, both along the coast because of sea-level rise and along rivers due to increased river flows. Major urban centers and ports such as St. Petersburg and Murmansk will face significant risks of flooding from extreme weather events.

Southern European Russia

Russia's traditional "breadbasket" region will face serious agricultural challenges as a result of drying and warming. Crop zones may shift and rural communities will face increased stress because of water shortages and possibly increased ethnic tensions. Considerable internal migration and migration from Russia's more severely affected southern neighbors may occur. Sea-level rise will be a long-term problem along the Black Sea coast.

The North Caucasus

One of the poorest and most ethnically diverse areas of Russia, the North Caucus is expected to become even hotter and drier. Already high unemployment rates will most likely grow, increasing the region's volatility and further pressuring internal economic migration to elsewhere in the Russian Federation.

Siberia

The most disruptive climate change impact in Siberia may be the large-scale melting of the permafrost, with attendant disruption to infrastructure. River basins will face a greater risk of flooding and the region will probably see significant migration from Central Asia.

The Russian Far East

Eastern Siberia and the Far East will become wetter, with risks of coastal and river flooding, particularly along the Lena River. The region may be the target of major Russian energy infrastructure expansion, which will be complicated by permafrost melting. Migration from Northeastern China will also probably continue, creating the possibility of a demographic sea change.

Overall Economic Challenges

The classic factors of economic growth for any country are the quantity and mainly the quality of its physical and human capital and the rate of technological change. So far in its transition from central planning to the market, post-Soviet Russia has been able to exploit its inheritance of physical and human capital. But this Soviet-era legacy is now reaching a point where serious deterioration of Russia's physical and human capital is a major obstacle to sustainable economic growth. It is almost certain that Russia will not be able to solve these economic challenges by 2030 and although it can survive and even prosper in the short to medium term without seriously addressing these challenges, by 2030 the costs of managing the annual impact of the physical and human capital deficiencies will be a major burden on Russia's economy.

In addition, Russia already faces a geographic challenge to its economic cohesiveness. If looked at in terms of "effective national territory"—defined as those parts of a state that contribute to the national economy at a level greater than that relative to its population— Russia resembles an archipelago of urban, economically dynamic islands amidst an ocean of vast, relatively undeveloped geographical space. Its effective national territory forms an "economic spine" that supports the rest of the country economically and strategically.

Moreover, the wealth that is largely generated by extractive industries across Russia's vast territory is concentrated in large state-sponsored monopolies in Moscow. Local populations and regional governments receive very little of their industries' profits and have very little say in corporate decisionmaking. This geographic disconnect encourages Russian industries to treat the resource areas they exploit and their inhabitants as disposable. It also removes industrialists from the proximate effects and impacts of climate change and gives them little incentive to be responsive in terms of adopting mitigation measures, leaving it instead to the under-resourced local governments.

Energy Challenges

Climate change has the potential for a material, negative impact on the single-greatest source of revenue to the Russian state—the oil and gas industry. In some respects, however, the effects of change may prove positive, such as in reduced energy consumption for heating, a major source of energy demand in Russia. Increased river flows may boost hydroelectric potential, and arctic melting may allow exploitation of previously inaccessible subsea hydrocarbon deposits.

The energy sector, particularly the revenue derived from hydrocarbon exports, is the most important factor that will affect Russia's near, mid-, and long-term future.[4] The market value of Russia's oil and gas resources—its resource rent—is primarily a function of how much Russia is able to produce and world market prices. Oil and gas rents are and will remain a far more important driver of Russian politics, economics, and foreign policy than climate change.

Although climate change will directly affect the production and transportation arenas of Russia's energy sector, the effect this will have in turn on resource rents is unclear. It does not follow that production levels or rents will fall because the total volume of Russia's oil and gas rents depends much more on price than on quantity. Between 1996 and 2007, the volume of oil produced increased by 63 percent. Nevertheless, price increases accounted for around 84 percent of total oil rent growth, while volume increase accounted for only 16 percent.

While production levels of oil and gas may slip in the years ahead as a result of climate change-induced impacts on energy infrastructure, the key question is how climate change may affect oil and gas prices, which have historically been more volatile than production levels. Presumably global supply of and demand for oil and gas—the fundamental determinants of price—will be affected by climate change, although the magnitude and direction of these effects could be quite uncertain. In addition, the global response to climate change will have a huge but still-indeterminate impact on demand for energy, particularly in light of decisions of the major economic powers regarding a move away from fossil fuels.

Oil and Gas Infrastructure and Development Challenges

The bulk of Russia's energy extraction infrastructure dates back to the Soviet period and therefore Russia was able to unleash tremendous excess capacity—and realize windfall revenues over the past decade—by applying more up-to-date techniques to the inefficient and wasteful legacy infrastructure. The gains in efficiency from this so-called "brownfield miracle" have now largely run their course, production levels—for example in Western Siberian oil and gas fields—are waning, and Russia will need to turn principally to "greenfield" development of new energy sources to keep production up. Many of the potential greenfield sources are in remote and exceptionally harsh locations in Russia's far northern and eastern regions. Vast new transportation infrastructure will be needed even to reach these areas, let alone transport the extracted energy to customers. Oil and gas extraction present roughly equivalent technical challenges and costs, and deposits are located in equally difficult geographic regions of Russia. Climatic challenges will further complicate this development, as permafrost melting and more volatile weather will impede access to and construction of vital new production areas.

In addition, Russia will face huge replacement costs for its Soviet infrastructure on top of the costs associated with new energy development. Climate change-induced effects, particularly permafrost melting, again are likely to add even more to these expenses because the Soviet-era infrastructure was built on the permafrost and designed accordingly—as the permafrost thaws earlier and deeper it will cause structural subsidence and force the replacement of a considerable amount of infrastructure that might otherwise simply have been repaired and updated. In contrast, the greenfield infrastructure can be designed from the outset to compensate for the altered conditions. Climate change may therefore render greenfield development, even in remote areas, more competitive with brownfield replacement than would otherwise be the case.

Regardless, Russia's development of the technically and climatically challenging new energy projects and replacement of its existing infrastructure will be the largest capital expenditure projects in the history of the planet. Russian sources estimate that developing oil and gas deposits in the Arctic regions will cost $500 billion, about the size of Russia's entire hard currency reserves at their peak in mid-2008. Not surprisingly, the current economic downturn has significantly reduced Russia's enthusiasm for launching these projects in the near term. While they have subsequently rebounded, oil and gas prices declined nearly threefold since the summer of 2008, dropping well below the $70 per barrel that the Russian Government considered necessary to make Arctic development feasible.

Challenges to Transportation Infrastructure

Climate change is anticipated to cause potentially serious periodic disruptions to Russian transportation and increase the strain on transportation infrastructure. As opposed to creeping environmental change in which effects occur slowly over time some climate- induced changes, such as flooding in major river basins including St. Petersburg, more snow in non-European Russia, and more frequent mudflows and landslides, will present immediate challenges to transportation and may separate Moscow, the economic center, from the regions and industries that fuel of the national economy. Such disruptions to commerce and economic activity may be sporadic and difficult to predict or to prepare for and will necessitate large-scale investments in infrastructure.

Permafrost melting is a major concern for the maintenance and viability for commodity flows of transportation systems, including those that serve the oil and gas pipelines in central and eastern Siberia. Because a significant proportion of road and railway networks are constructed on permafrost, increased melting could swallow the transportation networks of large expanses of Russia's territory into a swampy morass for a significant part of the year. Russia has a comparatively under-developed road and highway network and is far more dependent on railroads for long-distance transportation than most industrialized countries. Beyond the effects of permafrost melting on the rail system, increased river flows may threaten the structural integrity of railway bridges.

Waterborne transport faces mixed impacts from climate change. New ports and shipping routes in the Arctic may be offset by increased threats of flooding of existing ports and higher incidence of severe storms. Increased river flows may facilitate the use of rivers as transportation arteries, particularly in Siberia, but also brings increased risks of catastrophic flooding and ice jams. Additionally, the structures located at thousands of river crossings are unlikely to accommodate increased water levels associated with climate change.

Agricultural Challenges

The future impact of climate change on Russian agriculture is uncertain. Warming will lead to longer growing seasons, less frequent bitter winters, and more northerly areas open to cultivation. Expected resulting increases in yields will be at least partially mitigated by other climate change-induced effects such as greater weather variability, more frequent severe weather events, and the spread of plant diseases and pests into new areas.

The expected northward migration of planting zones will be a complex process with mixed impacts. The climatic conditions suitable for warm-climate crops such as cotton, grapes, tea, citrus, and other fruits and vegetables will move northward from Central Asia and the Trans-Caucasus into the North Caucasus and Volga regions. Depending on the specific crops, the northward shift may involve a broadening of the zone of potential cultivation. Alternatively, it may entail a shift in both northern and southern margins of the planting zone such that current cultivation areas will no longer be able to support historic crops. The latter scenario may have catastrophic effects on single-crop dependent agriculture such as in Central Asia. The outlook is further complicated by the fact that although climatic bands may shift, soil bands such as the Chernozem (Black Earth) will not. If climatic movement pushes crops into areas with unsuitable soil, yields may drop significantly and it may no longer be possible to cultivate certain crops at all.

Although new agricultural areas will open up, the areas on which Russia has historically depended and around which its agricultural infrastructure is organized will suffer increased stress. Many of the fertile Chernozem agricultural regions of southern European Russia—such as Belgorod, Voronezh, Kursk, Lipetsk, Orel, and Tambov Oblasts—are expected to experience water shortages and sustaining agriculture here will require increased irrigation and reliance on chemical additives. In the longer term, Russia faces a need to refocus its agricultural enterprises northward, which may entail uprooting an entire region's socioeconomic system.

Urban Challenges

As migration to urban areas increases, climate-induced hardships, such as the water shortages anticipated for the Moscow Oblast, will affect increasing numbers of people. Energy blackouts and brownouts resulting from unreliable or insufficient energy supplies can cause local economies and productivity to grind to a halt. Heat waves are likely to lead to increased levels of secondary pollution and public health stress. Water treatment in cities could be impaired and waterborne illnesses could rapidly affect an entire city. Infrastructure in many Russian cities is already decrepit and inefficient, and could be further damaged by climate change-induced effects.

Although overall climate change effects do not appear likely to create chronic or insurmountable disruptions for Russia's urban population, an increase in extreme weather events—including heat waves, storms, flooding, water shortages or contamination, and outbreaks of disease—could threaten cities with sudden local disasters.

Demographic Challenges

Unrelated to climate change but a factor certain to compound the impact of that change are demographic challenges that threaten Russia's human capital—that combination of the number and age of its citizens, their health, their education, and their location. A health crisis threatens productivity; a mismatch exists between the types of skills being produced by the educational system and economic needs; and mobility problems prevent reallocation of human resources in an effective manner.

A declining population is another aspect of this demographic challenge. The Russian Federation today is a great deal more homogeneous than was the Soviet Union but Russia's population has been in decline since the end of the Soviet Union. The Slavic population had been falling even before that time, although this drop was offset by high birth rates in the non-Slavic population. As a result, the Soviet Union was less than 50 percent Russian by 1989. Today the population is about 80 percent Russian, boosted by migration of ethnic Russians from the Former Soviet Republics since 1991.[5]

The population decline, although it has slowed in recent years, is primarily the result of a substantially higher mortality rate, particularly among working-age males, than other countries in Europe. Estimates of life expectancy by Russia's State Statistics Agency (Rosstat) suggest that after 2008 Russia will see the beginning of a decade-long precipitous decline in working-age population. By 2015, that group is projected to decline by 8 million and by 2025 by 18-19 million. This decline is more serious when one factors in the high numbers of the population who are incapacitated due to chronic health problems.

Given the downward demographic dynamics in Russia's Slavic population, increased immigration was already poised to play a greater role in filling Russian labor needs in the first half of the twentieth first century, leading to a more diverse nation with higher percentages of Asians and Muslims. Increased immigration because of climate change- induced stress in Central Asia and China can be expected to accelerate these trends. The prospects for further ethnic diversification as a result of immigration and disparities in birth rates have revived late Soviet-era fears that ethnic Russians may become a minority in Russia, although such fears do not appear to be borne out by the demographic facts. Popular conceptions of an internal Muslim demographic time bomb also seem exaggerated. There are less than 11 million Muslims in Russia, and their birth rates are dropping. Traditional Muslim birth rates among the larger Muslim ethnic groups such as Bashkirs and Tatars are not much higher than those among Slavs, and birth rates have even declined in the Caucasus.

Ecological Challenges

The effects of climate change will not be limited to human systems. Effects such as changes in growing seasons, permafrost melting, more frequent forest fires and floods, and the spread of disease will have broad-based impacts on Russian ecosystems. Entire ecosystems may shift into new geographic areas, such as the sub-Arctic taiga forest encroaching on more northerly areas previously dominated by tundra. In some parts of the North, warmer climatic conditions may lead to increased biodiversity. On the other hand, the stress associated with flora and fauna moving into new areas and competing with native plant and animal life may result in extinctions threaten and more species. The impact of climate change on Russia's vast wilderness ecology is particularly significant in part because many indigenous peoples depend on these ecosystems. Natural ecological change will in some cases be worsened by the secondary impact from human actions that are driven by increasing economic and climatic stress.

Climate change effects such as increasing sea-surface temperatures, changes in ocean circulation patterns, sea-level rise, changes in water levels in lakes and streams, and river and coastal silting caused by increased river flows, could create problems for fish populations. In the Caspian Sea, for example, an anticipated decrease in water levels is expected to reduce the number of sturgeon, an important industry in the region. It is not clear how climate change will ultimately affect fishing, but preliminary evidence suggests that in some cases fish schools are beginning to move to new waters, and in other cases fish stocks are decreasing.

In addition, climate change is likely to exacerbate the threat of disease caused by insects and vermin. Milder winters allow pests to thrive in more northerly areas, and Russia is already beginning to see an increased incidence of diseases more familiar in warmer regions.

Political Challenges

The sheer size of its territory and wide dispersion of its population have posed a perennial governance challenge for the Russian state. The country's extreme regional social, economic, and ethnic disparities are difficult to reconcile with the strong tendencies toward centralization of power and authority in Moscow. The oil center of Khanty-Mansiysk in western Siberia, for example, enjoys a per capita GDP comparable to that in the United States, while Vladikavkaz in the Caucasian republic of North Ossetia has a per capita income 30 times smaller. Moscow faces a huge political challenge in providing effective governance to cities and regions on both ends of this spectrum and every point in between.

During the Yeltsin years, state authority was diffused among the constituent regional and local levels of government, particularly to the ethnically defined republics within the Federation. Under Vladimir Putin, Moscow reversed this trend and undertook a sweeping recentralization of authority. The federal subjects—oblasts, republics, and krays—were grouped within seven large federal districts with centrally appointed governors. Local executives are heavily influenced by Moscow but locally elected. Although local parliamentary bodies are technically still elected, in practice higher authorities suggest candidates, whose election campaigns are state-funded. Power trickles down from the Federation level, curtailing the influence of budding, local-level civil society. Further erosions of federal subject authority may occur, such as the political restructuring of critical resource areas away from the control of the more sovereign federal subjects, such as the ethnic republics. This could be undertaken officially through boundary changes, or more insidiously by de facto economic relationships. The recentralization of authority already has aroused resentment among local actors who have sought to renegotiate their political rights and regain some of sovereignty from Moscow.

Such federal-local tension may be exacerbated if local authorities find themselves under increased stress because of climate change. Given Russia's geographic extent, regional climate change impacts will vary widely and may be best addressed at local and regional levels of government; but the state's tendency toward centralization of authority in Moscow may adversely affect the country's ability to respond to local climate change- induced challenges, worsening these problems and even causing them to spill over and affect the country more broadly.

CIVIL AND KEY INTEREST GROUP RESPONSES

Civil Society

The legacy of Soviet-era control over political and social activity continues to undercut civil society in Russia and the other former Soviet countries, both through lingering social inhibitions and continued state repression. Political and social activism has tended to be scrutinized, if not outright discouraged, by the government. As a result, few individuals will admit to involvement in a non-government organizations (NGOs), social organization, opposition political party, or civic group and most are cynical about what can be achieved through such channels.

Informal Network

Absent overt civil society mechanisms, many Russians have developed alternative, informal social networks to deal with day-to-day issues such as commodity shortages, which are likely to play a major role in how the public responds to the effects of climate change. The black market thrives and Russia is host to a large number of unofficial networks based on clan allegiance, ethnic identity, and any number of other criteria. These networks command a great deal of social power outside of the government, but are tolerated as long as they do not infringe upon the state's domain. Indeed, it is arguable that these networks actually strengthen the state: aided by the networks, the Russian people may feel they do not need to mobilize politically.

Ecological and Environmental Movements

Some civil organization nonetheless has occurred related to the environment and the environmental effects of climate change. The development of ecological movements, particularly in Siberia, has been increasing, bolstered by the interaction between these movements and indigenous cultural and ethnic preservation movements. Such collaboration is likely to increase as Russia feels greater effects from climate change.

In the Siberian North, where initial effects of climate change are being felt, indigenous activists are conducting their own environmental measurements because they distrust reports from the regional and central government. They have merged "folk knowledge" about climate change with broader scientific reporting and accelerated indigenous people's concerns. As a result, a high level of public awareness about climate change effects, such as methane outgassing from melted permafrost, now exists in this region. Public awareness in other parts of Russia is also likely to increase along with global awareness, particularly with the advent of more precise and reliable climate change measurements.

Local mobilization around environmental issues already has managed to convince Moscow to change course in some instances, as happened in the case of a pipeline proposed for the Lake Baikal area. More commonly, even when local and indigenous responses exceed what Moscow had expected, as in the case of a local referendum against a planned hydroelectric dam on the Lower Tunguska in Krasnoyarsk Kray, broader economic considerations remain the deciding factor. Even if local mobilization against mega-infrastructure projects such as dams and pipelines intensifies, it is not likely to rise to a level that seriously challenges state security or deters decisionmakers from pursuing national-level plans they perceive to be in Russia's overall interest.

Rural Responses

In general Russia's rural areas will bear the worst strains from climate change, including the immediate effects on agricultural productivity, interrupted connectivity to cities and sources of non-local commodities and energy, as well as health impacts of increased bogs, floods, disease vectors, altered temperature patterns, and food quality and supply. The ability of rural populations to cope will depend on several factors including the health of the local economy, the ongoing, toxic Soviet environmental legacy, and the degree of new climate change related stresses. However, rural residents may be more likely than then urban residents to have back up food supplies, traditional ways of dealing with sickness, and strong social networks.

The West and South

In considering how the Russian countryside will respond to climate change, it is important to distinguish between agricultural and non-agricultural rural populations. Russian agriculture—and its rural population—is concentrated in southern and central European Russia, from the Black Earth of the Ukraine to the southern Urals. Conditions in these areas vary: in southern Russia, the rural areas are more vibrant, with agricultural towns of 10,000-100,000 inhabitants who benefit from a favorable climate. The more northerly agricultural areas, in contrast, feature small, 19thcentury-style farms with low productivity and elderly populations.

These two markedly different rural agricultural models are likely to have quite different responses to climate change. Elderly rural dwellers are unlikely to move or generate conflict in the face of climatic stress, nor are state resources likely to be directed toward maintaining marginally productive agricultural areas. This way of life is already dying out, and climate change may simply accelerate the process. The response in the more intensive agricultural areas of the south is likely to be far more dynamic. Migration to the cities is likely to increase, as are demands for aid from the state, particularly in the area of expanding irrigation infrastructure, and conflict over water and other resources may become commonplace.

The East and North

The majority of Russia's territory, particularly in Siberia and the Far East, is a demographic and agricultural desert and will have a very different experience of climate change than the west and south. Russia's northern lands are a zone of poverty, soaring unemployment, industrial over-extension, and instability. The small and widely separated population is overwhelmingly urban and the tiny rural population— primarily indigenous peoples—is engaged in herding, hunting, fishing, and resource extraction, not agriculture.

The rural inhabitants of Siberia and northern European Russia have already begun to experience climate change-induced pressures, resulting in a loss of traditional means of subsistence and forcing many to move to increasingly larger towns and villages; however, any mass migration appears unlikely given the mixed impact climate change may have in these areas. Moreover, although shifts in traditional practices and ways of life will be traumatic, such shifts are already under way as a result of resource development and northern populations, already resilient to climatic challenges, are already responding to the leading edge of climate change.

Prospects for Rural Unrest

Historically, stress on rural areas was a major source of destabilizing political unrest in Russia but severe challenges to the state from the countryside are no longer plausible in post-Soviet, urbanized, industrial Russia. Rather, as noted previously, it is the lack of federal interest in or attention to their plight may be the most challenging aspect of rural society's response to climate change.

Urban Responses

Nearly three-quarters of Russia's population is urban. In some cities, the overall effect of climate change may be only a moderation of the harsh weather perhaps a welcome change for Russian urbanites. Given gradual climatic shifts, Russia is likely to have the resources to adapt its urban infrastructures to new conditions. Nonetheless, urban dwellers may face disruptions to transportation, food and water distribution, electricity, heating and air conditioning, waste removal and treatment, and public health. Such effects, perhaps severe in isolated cases, may be felt primarily by the very young, the very old, those with compromised immune systems, and those residing in hospitals or other institutions. Russians have low expectations regarding their government but climate change-induced challenges may shed more light on existing deficiencies and lead urbanites to demand more from their local and central government

Prospects for Urban Unrest

Russia's more recent historical experience suggests that the greatest danger from political unrest comes from the major cities, such as Moscow and St. Petersburg where the concentration of population and economic and political power means that a serious climatic event there could have major repercussions for the stability of Russia as a whole. The most serious threat may be a major flooding event in St. Petersburg. A truly catastrophic event of this magnitude in which federal and regional officials and agencies demonstrate tragic incompetence would be a massive blow for the state to absorb. Moscow is less threatened than St. Petersburg and other major cities. Owing to its privileged status, authorities will ensure the capital receives the resources it needs to cope.

Ethnic Minority Groups

Climate change may have a particularly significant impact on Russia's diverse ethnic minorities, depending on where they reside. Climatic shifts may disrupt traditional ways of life among groups—who are already under economic pressure—and in some cases force them to migrate and be subsumed within the larger Russian population and subjected to cultural dislocation, discrimination, and even violence. Climate change may spark increased resource competition among co-located ethnic groups. Groups living in isolated areas may suffer disproportionately from disruptions to the transportation network.

Moreover, climatic stress can be a key element inter-ethnic tensions devolving into violent conflict. For example, in northeastern Kenya in early 2009, drought and resulting crop failures sparked increased tribal clashes over water and pasture lands, high food prices, and the decimation of goats and sheep by a viral disease. Many minority groups in Russia live at a subsistence level in marginal areas where fairly minor climatic shifts could generate considerable local stress and resource conflicts that might play themselves out along ethnic lines.

The North Caucasus

The North Caucus region—the most ethnically heterogeneous region in the world and the site of numerous recent inter-ethnic or ethno-nationalist conflicts—often is cited as having the most potential for ethnic conflict as the result of climate change. As noted previously, while the region is projected to bear the brunt of climate change-induced drying and warming trends, it is one of Russia's poorest and least-prepared areas to cope.

However, when the North Caucasus is considered in aggregate, the region is extensive and expected climatic impacts in particular areas do not appear to correlate with existing ethnic fault-lines. Indeed, the regions expected to bear the brunt of water shortages and agricultural disruptions—the steppe and the Black Earth—do not have heterogeneous populations or unstable political situations; while areas characterized by social and political tensions—from Kabardino-Balkaria to Dagestan—appear more likely to experience agriculturally friendly trends such as increased water availability, more mild winters, and warmer weather. The North Caucus will certainly remain an ethnically charged region, but climate-induced conflict does not appear likely.

Siberian Indigenous Peoples

Indigenous peoples have shown a prescient grasp of the profound impacts climate change may have on their native regions. Siberians never needed scientists to tell them that their food chains were more fragile than other areas farther south. The lasting effects of marks from all-terrain vehicles and the easy destruction of, lichen, the main food source for reindeer, that rejuvenates in multi-year cycles or not at all, alerted them early after Russian mining and energy developers entered their territories. They have already seen and felt the loss of their hunting, fishing, cattle, horse and reindeer breeding territories in numerous places, at increasingly alarming rates. With projected climate changes they fear increased forest fires, oil pipeline fractures, and the return of long submerged illnesses—anthrax for animals, small pox for humans. Indigenous peoples, who believe in the delicate interrelationship of humans to nature, are already voicing their concerns over environmental degradation and climate change and tend to conceptualize climate change as nature taking revenge for human transgressions. These beliefs have manifested themselves in increasing interest in environmental issues and interaction with ecological groups.

To the extent that indigenous groups mobilize politically, they may become useful political levers for other interest groups. In the 1990s, ethnic Russian interest groups used indigenous mobilization as a way to leverage support for their own interests. In such situations, minority groups could become levers that challenge state authority even if their intrinsic political clout is minimal.

Internal Migration

The Russian Government's recently published climate change doctrine highlighted internal migration as a key concern arising from climate change. Russia already faces a great deal of economically driven internal migration, with movement from the East into the West, and from rural to urban areas.

Climate change will also "pull" migration into more northerly areas: the likely opening of new oil and gas fields in the north and east and the massive accompanying infrastructure projects will draw labor into previously sparsely-populated areas of Russia, as will the potential opening of Arctic shipping routes. Although large-scale shifts in population will no doubt cause social and economic strains, voluntary migration in response to labor demand is likely to be considerably less destabilizing than involuntary movement of refugees out of stricken areas.

However, large-scale, sudden or protracted internal migration because of climate change could create local distortions in the labor market, increasing competition, frustration, and racial violence. The urban metropolises of Moscow and St. Petersburg are already experiencing tensions and violence directed against Caucasian migrants and immigrants of Central Asian descent. Indeed, anti-immigrant hostility and xenophobia are deeply rooted and widespread on the streets of Russia and a growing number of violent neo-Nazi and others groups already blame minorities and immigrants for a host of economic and social problems.

The most destabilizing type of internal migration could be the displacement of agricultural populations into previously homogeneous rural areas and towns where they will compete for resources with established groups.

State Responses

While the Russian state will need to expend considerable effort to address climate change -induced challenges, historically it has successfully tackled large-scale challenges under harsh conditions. Given Russia's high overall state capacity and the mixed or comparatively tolerable nature of most anticipated climate change impacts over the next twenty years, climate change is unlikely to lead to a general failure of the Russian state.

The Russian political elite has historically viewed climate change as a net positive for Russia, anticipating that warming will alleviate some of the challenges associated with the country's cold climate. This attitude is now changing: some Russian leaders now realize the possibility that climate change may have a number of deleterious social, economic, and political effects on Russia, an awareness that is likely to lead to measures to alleviate some of the most dramatic impacts.

Climate change is likely to generate complications and challenges for the state in two principal areas. First, the differential impact of climate change across the Federation will complicate centralized governance and policymaking in Moscow. Russian authorities already face considerable challenges in managing a vast federal state with extreme regional socioeconomic disparities and considerable ethnic diversity. The addition of differing, and in some cases opposing, climatic shifts and consequent movement of people and economic activity will only make the Federation more difficult to govern.

Second, as previously noted, climate change will affect how the Russian state and its (largely state-sponsored) energy industries implement the massive infrastructure projects and investments needed for the continued development and solvency of the critical Russian oil and gas sector. These projects will have to be undertaken regardless of climate change, but climatic impacts may render the process more expensive and technically challenging.

Climate Change Mitigation Policies

In May 2009, in preparation for the Copenhagen Climate Conference later this year, Russia's Ministry of Natural Resources and Ecology unveiled a climate doctrine that outlines the country's response to climate change. The doctrine, prepared without public scrutiny, came as a surprise, particularly because of the change in tone among Russian officials that it reflects: rather than emphasizing the benefits of milder winters, the opening of the Arctic coast, and longer growing seasons, the doctrine warns of serious climate-induced challenges and outlines adaptive measures.

According to the Minister for Natural Resources and Ecology, Yuri Trutnev, Russia will need to take a strategic, long-term approach to adapt to climate change, looking at least to 2030 and as far as 2050 in some sectors. The climate doctrine is intended to provide guidelines for planning and instituting that policy approach, and calls for a range of initiatives:

- Increased research in order to acquire more precise and reliable data on climate change. Russia should establish a climate change oversight body and potentially a national center for climate studies.
- Expanded and updated state environmental regulation and legislation to bring Russia into line with international norms on climate change; more stringent regulation of industrial carbon and emissions;

measures to stimulate responsible resource use and efficiency; and better natural resource management.

- Structural change in Russia's economy aimed at increasing the resilience of key economic sectors such as agriculture, transport, and energy, to include a redistribution of resources and economic activity to meet shifts caused by climate change.
- Development of alternative energy sources to help meet domestic energy needs, as well as increased energy efficiency, an initiative in which Russia had not previously shown much interest.
- Participation in international climate change initiatives.

Although the doctrine reflects some new sensitivity on the part of the Russian Government to potential adverse consequences of climate change, it is not clear if the policies articulated in the doctrine go beyond rhetoric in part because the Moscow's support for the initiatives has to date been vague. In addition, critics have charged that the doctrine does not go far enough in addressing the root causes of climate change and rather focuses on alleviating the effects.

State Priorities

Except in wartime, few states have historically prioritized state security in as stark terms as Russia. The Russian Federation and its Soviet precursor have been willing to sacrifice progress in virtually every social, economic, and technological area in order to focus on security and the military. Even today, with an emphasis on energy export-led economics, state security remains the ultimate priority of Russia's decisionmaking elite.

Russia's approach to state security is characterized by caution and paranoia. Having weathered a huge economic and political crisis in the 1990s, Russian leaders are very conservative and place a premium on precautionary measures. Moscow feels a sense of vulnerability because of its relative demographic and geopolitical decline vis-à-vis both its longstanding rivals and new, rising powers. As a result, Russian planners and decisionmakers typically focus on worst case scenarios and remain largely unwilling to depend on other countries for food, energy, or physical security.

A state whose primary goal is security will respond quite differently to climate change effects than one oriented toward economic efficiency. The goal of efficiency dictates policies such as competition, pluralism, entrepreneurial independence, and participation in the global division of labor. An emphasis on security, in contrast, leads to excessive control, a high degree of redundancy, lack of specialization, resistance to interdependence and openness, and a focus on building up reserves of various kinds. Climate change- related tradeoffs will be framed in terms of their net effect on state resilience more so than their economic impact. Climate change-induced challenges are likely to necessitate a broad range of state interventions including mega-infrastructure projects and local-level responses to discrete incidents.

To insure that these challenges do not rise to levels that threaten security, the state must bolster its overall capacity to respond and adapt. Therefore, if expenditures are seen as preserving state power and stability, the Russian state may be willing to spend more on climate change mitigation than a growth-oriented state. Although Russia favors efficiency and global cooperation in responding to climate change, Russian leaders, preferring autarkic solutions, will most likely not allow the country to be placed in a dependent position.

Policy Decisionmaking

The current Russian leadership has tended strongly toward a pragmatic approach to policymaking aimed at keeping Moscow's options open until the outcome is fairly certain. The Russian Federation is much less bureaucratic than the Soviet Union; decisionmaking is more personal and less institutionalized; planning is less complex and deliberative; and the policymaking process is opaque and in many cases a formal process may not exist. Another critical feature of decisionmaking in Russia is the conflation of Kremlin officialdom and the corporate elite, to the point that the distinction between state policy and corporate policy is often arbitrary.

Overall, the Russian state leadership expends much less effort to monopolize decisionmaking than was the case in Soviet era, primarily stepping in if an issue is deemed a national security threat and/or requires a major state initiative to address it. In such cases, both the Soviet and post-Soviet regimes have shown a capacity to rapidly come to decisions and act on them, which may facilitate the rapid adoption and implementation of major initiatives to deal with climate change related incidents. But such actions will tend to be reactionary and done with an abbreviated policy formulation process that has in the past led to deeply ill-considered policies with disastrous long-term side effects.

State Capacity

Russia faces a very complex planning, prioritization, and implementation challenge regarding its responses to climate change. Nevertheless, heavily industrialized and well- educated, with an outsized national security establishment, the Russian state already has a robust capacity to respond to climate change, which could be boosted even more with investments over the coming years.

Analysis & Forecasting

Russia has a strong scientific establishment and its research institutes and universities provide ample expertise and infrastructure with which to gather data on climate change, analyze it, and produce forecasts. The lead agency involved in such efforts is the Federal Service for Hydrometeorology and Environmental Monitoring (Roshydromet), the Russian equivalent of the US National Oceanographic and Atmospheric Administration (NOAA).

Emergency Response

Russian capabilities in emergency response and disaster consequence management are relatively robust. The lead agency for emergency response is the Ministry for Affairs of Civil Defense, Emergency Situations and Disaster Relief— better known as the Ministry of Emergency Situations or the Emergency Control Ministry (EMERCOM)—the Russian equivalent of the US Federal Emergency Management Agency (FEMA). EMERCOM has a reputation as a dynamic and effective agency, and has capabilities to respond to a wide variety of climatic incidents throughout the Federation, including forest fires, floods, and storms. Acting through the Prime Minister's office, EMERCOM can draw on assistance from the Ministry of Defense and Ministry of the Interior, as well as assistance from foreign and international disaster response agencies.

Environmental Protection

In contrast to emergency response, environmental planning and prevention are weak. Environmental oversight within the Russian state bureaucracy is housed in the Ministries of Regional Development and of Natural Resources and Ecology, both of which are primarily charged with facilitating economic development and resource extraction. These ministries collaborate with the increasingly compliant, non-elected republic and regional governments. This creates a conflict of interest situation in which development goals supersede environmental concerns and ecological oversight is pro forma or corrupted. Problems include data collection, secrecy, and, especially, a lack of systematic efforts to curtail or repair environmental harm with sustainable development.

State Resources

As mentioned earlier, Russia's oil and gas wealth is by far the most essential element of its "adaptive capacity." A historical correlation exists between the value of oil and gas rents, on the one hand, and Russian and Soviet economic performance and political behavior, on the other (Figure 1). When resource rents are high, the Russian political system tends to be assertive, centralized, and less prone to reform. When the rents grew in the 1970s and early 1980s, the Soviet Union entered an expansive phase both economically and politically. The subsequent collapse of oil rents was followed in short order by the collapse of the Soviet Union. The sharp growth of oil and gas rents in the late 1 990s-2000s correlates with the resurgence of authority and stability for which Vladimir Putin is often given credit.

The future price of oil (and gas, which follows the oil price closely) is one of the most uncertain of all economic parameters. Taking a high-price scenario of oil prices averaging $120 a barrel through 2030 and a low-price scenario of $40 a barrel, forecasts of Russian oil and gas output can be used to give a plausible range for Russia's future resource rents. In either case, rents would provide Russia with a great deal of wealth to meet significant climate and aging infrastructure challenges if they can be efficiently and effectively applied to that end. In the high-price scenario, Russia would enjoy a staggering amount of wealth over the next 20 years, on average 30 percent higher than in the historically boom year of 2008. Even under the low-price scenario, the average annual rents would be 80 percent higher than in the preceding 21-year period, 1987-2008. Only the unlikely combination of an extended period of very low oil prices and poor policy in managing rents will produce a Russia as poor and weak as it was in the 1 990s.

State Repression

When considering whether instability generated by climate change might pose a threat to state security, it must be taken into account that the Russian Federation has an extensive and capable repressive internal security apparatus at its disposal. The Federal Security Service (FSB) is a pervasive presence that has largely reclaimed the virtually unchecked powers it had in its previous incarnation as the Committee for State Security (KGB). The Ministry of the

Interior (MVD) disposes of some 200,000 fully equipped internal security troops with which to put down armed insurrections. Special Purpose Police Squad (OMON) riot control units are stationed in every Russian oblast and major city.

Civil groups and non-governmental organizations are allowed, but are continually scrutinized, discouraged, and hindered in their activities in many indirect ways. Such coercive measures may be directed at ecological movements that might otherwise assist the state and society in responding to climate change-induced challenges. State coercion already has included publicly discrediting leading ecological figures and permitting or directing attacks on ecological groups by nationalist extremists.

"Level" of the Response

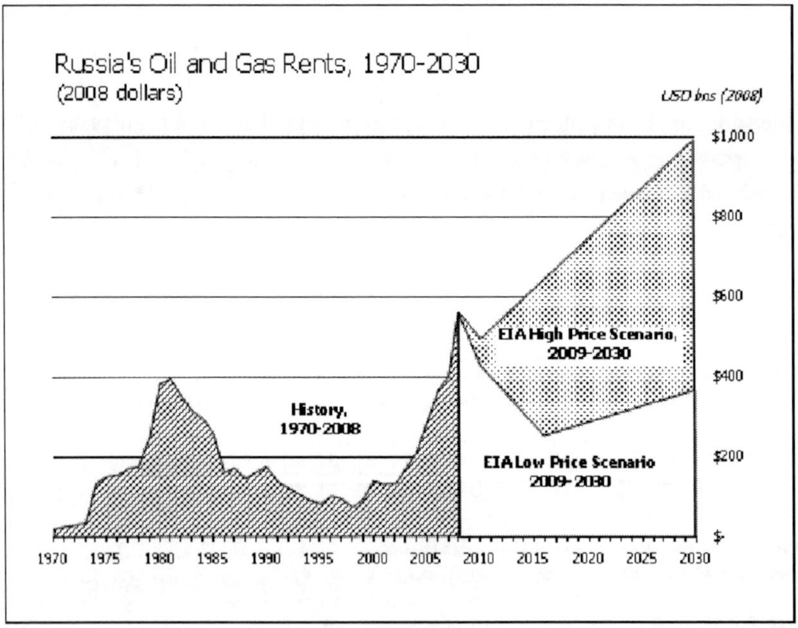

Figure 1. Russia's Oil and Gas Rents, 1970-2030. Source: Clifford G. Gaddy, data from US Energy Information Administration (EIA), "International Energy Outlook 2008."

The governments of the constituent republics, oblasts, and krays in many respects will be better positioned to observe and respond to the considerable variety of climate change- induced challenges across Russia. Horizontal mobilization across republics, oblasts, and krays is possible—perhaps organized around shared climate change impacts. State centralization, however, has robbed the federal subject governments of authority and resources, which may dramatically reduce their effectiveness as first responders to the effects of climate change.

Ineffective policy responses from Moscow, lack of consideration of local or regional interests, and lack of attention to local or regional problems arising from climate change may lend increased impetus to demands for decentralization from the constituent levels of the Federation government. Climate change may thus become a rationale that regional and local governments can use in negotiations with Moscow for greater authority and resources.

Territorial Integrity

Russia is set to undergo a geographic shift in economic activity and population unrelated to climate change in which regions distant from Moscow may experience significant gains in population, development, and economic power. If coupled with a perception that Moscow is unresponsive to local concerns—such as local effects of climate change—this increased clout may cause governments in the peripheral federal subjects to question their allegiance to the Federation. For example, if Sakhalin Oblast, in the Russian Far East, north of Japan, becomes a key energy producer, its leaders might question how it serves their interests to continue to be integrated with Russia rather than pursuing closer ties to their proximate customers in Northeast Asia.

That being said, the central government is highly unlikely to allow threats to the Federation's territorial cohesion to fester, as the precedent of Chechnya illustrates. The reassertion of Federal power under Vladimir Putin has for the time being reined in the federal subject governments, and if they try to reassert themselves over the next two decades they will have a steep hill to climb. In addition, Moscow has shown interest in creating more cohesive national infrastructure as a means of strengthening ties with distant regions, and is incorporating that strategic intent into its development projects. The Eastern Gas Program, for example, envisions investment in gas-dependent industries (such as fertilizer production) in the Russian Far East and the connection of the Far Eastern gas fields with fields as far west as Lake Baikal. These efforts may be driven by a desire to reinforce cohesion more than economic considerations: if Moscow is able to maintain political influence over its distant territories in the Russian Far East, resource development in the region could serve as a catalyst to hasten the integration of Russia with the economies of the Pacific Rim.

Urban Planning

Climate change-related problems in cities have the potential to rapidly become severely destabilizing. To the extent that municipal governments remain underresourced and subject to cumbersome, overly centralized state control, they have the potential to become weak links in Russia's state response capacity.

Environmental Protection

The federal subject governments are much closer to the ecological issues facing their regions than the central government and are more responsive to local ecological activism. The federal subjects of the Siberian North, strongly influenced by the beliefs and traditions of their indigenous inhabitants, have been particularly active in autonomous environmental preservation measures. The Sakha Republic, for example, features an excellent network of ecological protection zones: two major federation level preserves, five national parks, 78 reserves, and numerous other protected areas of local interest. The Sakha Republic Ministry for the Protection of Nature publishes yearly reports available on the Internet documenting at least some efforts at monitoring ecological devastation and planning for more Sakha preserves in the face of huge pressures for further development of the republic. Nevertheless, local authorities are sometimes loath to admit their failure to protect the health of their communities by publishing data on environmental degradation and pollution. In addition, local authorities have in many cases allowed mineral and energy exploitation to become priorities, overriding local interests even while giving lip service to the importance of ecology and "lands of traditional means of subsistence."

State Economic Policies

The former Soviet regime managed its resources and economy poorly and fell so deeply in debt to Western governments that it sacrificed its financial, and in effect its political, sovereignty. From this experience, Vladimir Putin and his associates concluded that financial and fiscal health were essential. Consequently, as oil prices rose, the Putin regime prioritized the use of the windfall to pay off the country's foreign debt and build reserves. Russia's foreign exchange reserves, now the third largest in the world, play a critical role primarily in protecting Russia's financial sovereignty and secondarily ensuring the welfare of its citizens. Looming increases in expenditures and greater economic uncertainty as a result of climate change are apt to reinforce these tendencies.

Marginally but not dramatically lower levels of revenue derived from hydrocarbon exports may also spur more rapid, but still limited diversification of the Russian economy. Public rhetoric in favor of "diversification away from oil" notwithstanding, Moscow is not likely to diversify away from oil and gas despite its vulnerability to the vicissitudes of commodity prices.

Energy Policies

Russia enjoys a major comparative advantage in hydrocarbon-based energy production that will persist regardless of increased production costs associated with climate change. Indeed, for the foreseeable future, Russia will remain reliant on fossil fuel commodities and its energy planning envisions only modest diversification away from hydrocarbons, mainly as an excuse to subsidize particular alternative energy industries. To the extent that diversification to supplement oil and gas does occur, hydroelectricity, nuclear power, and liquefied natural gas are the most likely choices.

The Russian Government nevertheless is showing interest in energy efficiency for the first time. A recent World Bank study concludes that Russia can save up to 45 percent (equivalent to nearly six million barrels of oil per day) of its total primary energy consumption by adopting measures that could pay for themselves within four years,[6] putting Russia in an excellent position to address any increase in demand or decline in production as a result of climate change. In addition, Russia has announced the intention to bring domestic energy prices up to international export parity by 2011. To the extent that Russia's domestic energy demand is elastic, this measure should reduce energy consumption, which is inflated by artificially low domestic energy prices.

A focus on increasing efficiency may delay implementation of the necessary energy- infrastructure improvement programs, but it cannot substitute for them. Indeed, as noted previously the major focus of Russia's domestic energy policy will be the vast capital expenditures associated with replacing the aging Soviet-era infrastructure that forms the core of Russia's energy production enterprises. Russia and its investment partners will need high prices in order to lay out the initial investments and will also need to be able to count on sustained high prices over the considerable period it will take to bring greenfield energy infrastructure online. Developing oil fields in eastern Siberia will require $70 a barrel oil prices to be a profitable venture.

Russia's simultaneous need to replace its aging infrastructure and to modify its infrastructure to take climate change into account is in some respects fortuitous. For example, an important preventive measure to reduce the chance of spills if climate change renders the permafrost-based pipeline systems unstable would be to elevate pipeline infrastructure and deepen its physical foundation to minimize risks of subsidence. As long as Russian infrastructure planning incorporates those changes necessary to cope with such climate change effects as permafrost melting, or exploit opportunities, such as the opening of new ports in the Arctic, Russia will only have to invest once to address both issues. Greenfield development in the Barents Sea and Far East will require a great deal of new construction projects. Historical infrastructure, much of which will need replacement, will be less important in supporting most of the greenfield projects, although some will still rely on existing networks. Development areas such as the Yamal Peninsula, for example, will require incremental new construction but will be connected to existing infrastructures.

Russia will face an increasing incentive to accept foreign involvement in these expensive and risky energy development projects, at least on the financing end of the project. Extending the existing pipeline infrastructure to reach the proposed Yamal field will cost $10 billion, even before considering costs of field development. The economic incentives pressing Russia in this to involve foreign investors nevertheless be filtered through a security lens by Kremlin decisionmakers and much of the critical supporting infrastructure—roads, railroads, bridges, ports, new settlements—will be paid for by the Russian state.

Lastly, these large-scale infrastructure projects undoubtedly will be accompanied by waste, corruption, environmental degradation, and outright construction errors. Assuming errors are even fixed, repairs and replacements will require all the more capital outlays. For example, a new oil pipeline being built near the Lena River in the Sakha Republic is being built as a trench, rather than a tunnel, with increasing possibilities of oil spills in conditions of melting permafrost, even though the necessity of surrounding pipelines with tunnels or building them aboveground is obvious to locals in the area.

Hydroelectricity

Increased river flows are expected to substantially boost Russia's existing hydroelectric capacity and hydropower potential but, as with the rest of the energy sector, the hydroelectric infrastructure will require substantial investment. Whereas in many countries, notably China and India, hydropower development is an international flashpoint, this is unlikely to be the case in Russia. Although most of its major rivers run almost entirely within Russia's borders, eliminating the common international issue of diversion by downstream countries, Russia may be less likely to pursue hydropower because oil and gas development will absorb most of the infrastructure investment.

Nuclear Power

The nuclear power industry currently supplies 16 percent of Russia's electricity needs, and Russia plans to produce at least a quarter of its domestic electricity through nuclear power by 2030. The industry is managed by Atomenergoprom, a civilian nuclear power monopoly subordinate to the Russian State Corporation for Atomic Energy (Rosatom). Atomenergoprom includes Energoatom, which operates Russia's civilian reactors; the nuclear fuel producer and supplier TVEL; nuclear goods and services exporter Tenex; and nuclear facilities builder Atomstroyexport. Although Atomenergoprom has ambitious plans to expand its export activity, international concerns about proliferation make nuclear exports to some countries even more politicized than gas pipelines. The nuclear industry, like oil and gas, needs to replace aging Soviet-era infrastructure; however, in the event of a prolonged global financial and domestic budgetary crisis, the nuclear industry will have to compete with the far more profitable hydrocarbon industries for state funding and political cover, which may constrain nuclear power expansion. In order to fund domestic infrastructure needs, Russia may turn to increased exports to such customers as Iran in order to fund domestic infrastructure needs, with corresponding implications for global security.

Liquefied Natural Gas

As Russia moves to develop new natural gas fields, it has the option of diversifying from pipeline-based gas delivery into liquefied natural gas (LNG). Pipelines create a hardwired relationship between Russia and its customers, primarily in Europe, and increasingly in China. Pipelines require consistent demand to be profitable, and cannot easily be redirected to new customers. Pipeline-delivered natural gas is therefore less commoditized than oil, which is bought and sold on the world market rather than being directly delivered to specific customers. In contrast, tanker-shipped LNG allows more flexibility; the LNG can be shipped to different customers as demand shifts. This flexibility may become more appealing to Russian leaders and to the gas monopoly Gazprom if they assess that some customers, particularly in Europe, may try to move away from a fossil fuel-based economy. On the other hand, a less hardwired natural gas sector is likely to increase natural gas price volatility, making Russia's economy more vulnerable to external shocks.

Gazprom is likely to add LNG to its repertoire while retaining pipelines as its principal focus. Historically Gazprom has avoided LNG in order to protect its pipeline monopoly, but it has entered the LNG business with its participation in the Sakhalin-2 natural gas development project in the Russian Far East. The firm also has drawn on foreign contractor expertise in LNG and deep-water gas drilling, as in the case of the Shtokman field in the Barents Sea. Nevertheless, incremental updating and expansion of the existing pipeline-based natural gas export system remains cheaper than development of a greenfield liquefaction capability.

Water Policies

As discussed previously, Russia faces major technical and engineering challenges to manage the impact of hydrologic changes and soil subsidence from melting ice, but the country has long battled harsh climatic and geological conditions with considerable perseverance and ingenuity. Russia's water supply is expected to increase overall, but much of the increase will occur in sparsely populated areas such as Siberia. Indeed, the differential climate change impact on water availability across Russia raises the question of whether the state will turn to large-scale water redistribution. Water will be relatively overabundant in the north, while the south, where Russia's key agricultural areas and most turbulent regions are located, will face shortages. Dating back to the Soviet period, Russian leaders have considered reversing the Siberian rivers to flow south, rather than unused into the Arctic Ocean. Such grandiose civil engineering projects may be fanciful, but more reasonable projects to transfer water from swollen northern rivers to the drying southern cities and fields are plausible. The expected major increase in flow from the Volga River, for example, could supply irrigation to the adjacent steppe regions expected to face water shortages.

Russia will most likely need to build new reservoirs, dams, and water pipelines to adapt to a new hydrological environment. Such projects would have to compete for resources with the equally massive infrastructure investments required in the energy sector.

Transportation Infrastructure

At the same time, the Russian state faces a large-scale requirement to upgrade the country's transportation network to adapt to changing climatic conditions. Many of Russia's roads, railways, and bridges are constructed on permafrost and its melting is likely to necessitate laborious improvement of thousands of miles of roads and railways in northern Russia, arteries needed for the expansion of infrastructure in the North. Potential competing focuses for investment include a more substantial snow removal system, ways to prevent ice-jamming of major rivers, flood control systems, and fortified bridges and river crossings. Much of these infrastructure improvements will have to be undertaken before substantial progress can be made on the all-important energy infrastructure issue. Accordingly, portions of the transportation network associated with the energy sector will no doubt receive priority, potentially leaving other parts of the country and their inhabitants with damaged or inadequate access to transportation for an extended period.

Regional Implications

Russia not only enjoys a better climatic outlook than many of its neighbors, but it has comparatively greater ability to cope with climate change. As a result, the most daunting climate change-induced challenges Russia will face out to 2030 may originate from outside its borders.

Regional Energy Issues

Climate change is likely to affect Russia's relations with its regional energy clients in two principal respects. Firstly, climatic stress on Russia's pipeline infrastructure will likely lead to increased maintenance costs and increased development costs for new oil and gas fields. Russia probably will pass these costs on to its customers in Europe and East Asia. Secondly, Europe is likely to forge ahead with efforts to transition away from fossil fuels and to alternative sources of energy in the interest of reducing emissions and mitigating longer-term climate change. This would jeopardize demand from Russia's primary energy customer.

Dynamics between Russia and its European energy customers are troubled. Russia shut down its gas supplies to Ukrainian transit routes in January 2009, and the situation escalated into a gas blockade affecting much of Europe. Gas supplies were shut off for eighteen days until Russia and Ukraine finally signed a memorandum of agreement. Russia's leaders may have been hoping to demonstrate Russia's critical role in European energy security and to persuade European countries to fund new pipelines that would increase European access to Russia's gas supplies. But Russia could face a shortfall in its gas production once European demand recovers if Russia does not invest in new gas field development. President Medvedev underscored the importance of constructing the proposed Nord Stream pipeline on the Baltic Sea bed to provide direct gas links to Germany and the South Stream pipeline that would bypass Ukraine, carrying Russian gas exports over the Black Sea bed to the Balkans and Central Europe. In addition, Prime Minister Putin promoted Yamal Two, the proposed pipeline westward through Belarus and Poland, and the expansion of Blue Stream, by way of the Black Sea from Russia to Turkey, as well as a proposed gas pipeline eastward to China that would parallel a planned oil pipeline. Some question whether or not Russia has the gas resources to fill these pipelines, however.

The European Union has not responded to the gas crisis in an integrated manner by investing in diversified pipeline projects with Russia, although some individual European states have made such investments. Rather, the EU has launched a serious effort to find non-Russian sources of natural gas, such as the proposed Nabucco pipeline which would allow access to Caspian gas via Turkey, bypassing Russia. Because pipelines represent a major investment and a long-term relationship, the reliability of the supplier is a critical consideration and Nabucco does not have sufficient non-Russian supply to be convincingly viable. Russia's provocative use of its gas pipelines as economic "weapons" may have branded it an unreliable supplier with serious long-term consequences for the country's natural gas industry and have put European investment across Russia's energy sector in some jeopardy. Without European investment, the proposed improvements to Russia's export capacity would take longer to bring on line.

In addition to oil and gas, Russia exports electricity to Scandinavia, Ukraine, and the Baltic States, and imports it from the Trans-Caucasus. Expected gains in Russia's hydroelectric potential as a result of climate change raise the possibility that electricity exports could be expanded. At the same time, greater risks to transmission lines from soil subsidence and stronger winds are likely to complicate long-distance electricity transmission schemes.

The Arctic Ocean

In contrast to their conservative attitude toward other aspects of climate change, Russian scientists and officials have paid keen attention to the dramatic developments associated with rising Arctic temperatures and melting ice caps that have accelerated in the last several years. Having by far the longest contiguous Arctic coastline and a preponderance of ice-capable naval capabilities, Russia is ideally positioned to exploit the opening of the Arctic Ocean as an arena for trade and resource exploitation.

A key Russian Federation Security Council policy document entitled "On Defense of the Russian Federation's National Interests in the Arctic," frames Russia's goals in the Arctic, stating [7] that "the Arctic must become Russia's main resource base," and lists "active development of natural resources" as the first priority of the Russian Federation in the region. It calls for "international recognition of the outer boundaries of Russia's continental shelf in the Arctic," an oblique reference to Russia's claims of the Lomonosov and Mendeleev Ridges that would give Russia jurisdiction over half of the Arctic Ocean. Russia claims that these subsea ridges extending far out into the Arctic Ocean are linked to the continental shelf and thus part of Russia's legitimate maritime territory. Russia claims nearly a half million square miles of seabed beyond its recognized 200-nautical mile exclusive economic zone.

Direct conflict over Arctic resources may be unlikely, since the majority of the energy reserves in the Arctic are in uncontested areas. The incentives for cooperation—both in resource extraction and support to Arctic economic activities—may be greater than has been conventional wisdom. In the limited areas that are contested, states that anticipate exploiting resources have already begun to maneuver in support of their claims. Denmark, which chairs the Arctic Council for the next two years, has made clear that it considers securing Arctic resources a priority. Negotiations are complicated and Russia is sensitive to the fact that all the parties in disputes over Arctic territory except

Russia are NATO members (the United States, Canada, Denmark, and Norway).

Beyond the Arctic's potential energy resources, the melting of the polar ice cap has major international implications, specifically for maritime transit. Formerly impassible sea lanes, such as Canada's fabled Northwest Passage, are already beginning to open to shipping as the pack ice in the Arctic Ocean recedes. The opening of the so-called Northern Sea Route (NSR) along Russia's Arctic coast promises to cut thousands of miles off the shipping routes between Europe, East Asia, and the North American Pacific coast. These shipping routes will create new strategic interests on Russia's northern periphery for Russia, NATO, and non-contiguous countries anticipating heavy reliance on the NSR, particularly China and the EU.

The sea lanes of the Arctic Ocean will remain perilous: rising water temperatures will increase the power and severity of Arctic storms, which along with icebergs will potentially hamper shipping and the exploitation of offshore energy deposits. Russia is likely to play a major role in search-and-rescue (SAR) and weather monitoring for Arctic shipping in the rough northern seas, which may facilitate a more cooperative approach among the Arctic states. For example, the US Coast Guard already works well with the Russians in the Arctic.

The receding ice also opens up potential ports along the Russian northern coast, which could allow Russia to more easily export onshore energy from northern Siberia. Historically, Russia's main strategic challenge was to secure access to warm water ports for both military and economic purposes. In effect, climate change will bring the ice-free ports to Russia, removing one of the most durable strategic rationales for Russian expansionism. Russia is already building a new class of winterized, ice-capable tankers to transport oil in rough Arctic waters.

New ports on Russia's Arctic coast may provide vital shelter to ships during violent storms along the NSR, along with SAR and emergency infrastructure, and refueling stations. Although Russia's fleet of Arctic-capable icebreakers may be less required given the receding ice, the ships may assume a SAR role as shipping traffic increases. Lastly, revitalization of Arctic ports may be a means to extend state administrative control into previously isolated areas of Russia's North. To date, Russia's commitment to developing its Siberian ports has been mainly rhetorical, but this is likely to change as the NSR opens to shipping traffic.

Russian enthusiasm about the Arctic surged at the time of rapid economic growth and high oil prices, peaking in 2007 with the planting of a Russian flag on the seabed beneath the North Pole. The current global economic downturn significantly reduces the incentives for Arctic development, both onshore and offshore. Observations of Russian media indicate that the "Arctic buzz" has waned since the onset of the economic crisis in the fall of 2008. Nevertheless, Russia's sense of entitlement to the Arctic's resources and the likely continued opening of the ocean because of melting ice ensure that the "buzz" will be heard again.

Fishing Conflicts

Climate change may generate increased disputes over fishing rights in the Arctic and Russia's peripheral seas. Changes in the movement of fish schools could bring Russian and other fishermen into conflict over new fishing grounds and exacerbate competition in existing grounds, particularly in the Arctic Ocean and the Baltic Sea. While unlikely to drive international conflict at the strategic level, particularly when compared to the subsea resource issue, fishing conflicts have great potential to generate local incidents that raise the overall level of tension in the seas along Russia's periphery.

Disputes over fishing rights and the relocation of fishing grounds could become a larger factor in relations between the Arctic states, particularly Russia and Norway, which have had disputed control of waters—in terms of sub-sea resources and fishing rights—in the western Barents Sea for decades.

Cross-Border Migration

Immigration into Russia currently follows a "pull" scenario: migrants enter Russia pulled by better employment opportunities. In European Russia and Siberia, the majority of migrants come from Central Asia and the Trans-Caucasus (Armenia, Azerbaijan, and Georgia), while temporary Chinese laborers are a major presence in the Far East. Climate change-induced migration would instead follow a "push" scenario, in which adverse climatic effects in neighboring countries would drive migration of environmental or economic refugees into Russia. Water shortages and more frequent droughts, particularly in Central Asia, Mongolia, and Northeast Chin, could push large numbers of people into Russia and become a source of instability, especially in hard economic times. As with internal migration, however, it is not clear that populations under climate- induced stress will respond by migrating rather than taking measures to adapt.

Moreover, any climate change-induced migration into Russia may in fact be less destabilizing than previously considered, as an increase in "push" migration may coincide with a decline in "pull" migration. Much of the immigration into Russia has been essentially a redistribution of population after the collapse of the empire. A large proportion of migrants have been ethnic Russians residing in the former Soviet Republics, but that source of migration is now almost depleted. The attractive influence of social networks and economic incentives may also be reaching a saturation point. In the medium term, as "push" migration becomes a larger factor, Russia will begin experiencing the previously mentioned expected steep decline in working-age population, and the demand for labor will increase. Russia might consider a guest worker program similar to those in Europe.

Migration from Central Asia & Mongolia

Climate change may have an especially severe effect on conditions in arid Central Asia, particularly in terms of water shortages and disruption of agricultural patterns; but a number of factors, including historical precedent, suggest that a rapid and destabilizing exodus from Central Asia into Russia is unlikely. Environmental conditions in Central Asia—such as air contamination of large swaths of Kazakhstan and Uzbekistan resulting from the drying out of the Aral Sea— degraded significantly in the late 1980s and continued to deteriorate rapidly in the 1990s. Although these adverse conditions affected all ethnic populations in the region, the overwhelming majority of migrants to Russia were ethnic Russians, as by and large, non- Russian groups remained in their home countries. Moreover, most large population centers in Central Asia lie far from Russia's borders in the southern parts of the region, such as the Fergana Valley. Iran, South Asia, and even western China are closer potential destinations for migration than Russia. Passenger transit is limited and costly, and the Russian Border Guard Service—now part of the FSB—mans border checkpoints at the main road and rail entry points from Kazakhstan into Russia.

Although it may face similar climate change-induced pressures to the Central Asian states, Mongolia has too small a population to pose a migration challenge to Russia. If migration from Mongolia increased, the ethnic Mongolian populations on the Russian side of the border would probably help absorb the newcomers. Moreover, "push" migration out of Mongolia is more likely to be directed toward China.

Migration from the Trans-Caucasus

Migration from Azerbaijan, Georgia, and Armenia may prove particularly troublesome given the existing ethnic tensions. Bitter memories persist of the Russian apartment bombings in 1999 that were blamed on people from the Caucasus and contributed to the intensification of Russia's military engagement in Chechnya. To date, migrants from the Trans-Caucasus have preferred to settle in the large cities of European Russia where they can find work rather than in the North Caucasus where conditions are similar what they seek to escape.

Chinese Migration

The numbers of Chinese economic migrants crossing the border into the Russian Far East are already considerable and increased migration could create tension between the two countries, eroding recent improvements in bilateral relations. Unlike the Central Asian states, however, China ranks high in terms of resilience to climate change, so "push" migration may be less of a factor and the Chinese state, with its increasing adaptive capacity, is likely to be able to act to cushion adverse climate change impacts. Nevertheless, sustained climatic pressure from desertification in China's Northeast could still increase Chinese migration into the Amur River valley and Russia's Maritime Province (Primorskiy Kray), areas where the ethnic Russian population is in decline.

Moreover, labor shortages in the Russian Far East will continue to drive "pull" migration from China, although climate change may somewhat alter the pattern of this migration. The southern areas of permafrost, now near China's border, are likely to move northward 90-120 miles, which may negatively affect agriculture in the Amur River region because of soil deterioration and more pests and make the area less attractive to Chinese agricultural migrants. At the same time, Arctic warming could shift the northern borders of the forests and steppes north by up to 600 miles and tundra areas become more hospitable. In the longer term this shift could open more of the interior to the timber industry, which already employs many Chinese migrants in Primorskiy Kray and other parts of the Russian Far East. These shifts could gradually induce more Chinese migrants to penetrate further into the Russian interior rather than clustering near the border.

A real potential exists for the Russian Far East to become demographically Chinese, although driven more by an outflow of Russians than an influx of Chinese. Although the Far Eastern Federal District comprises over a third of Russia's territory, less than six million people live there, concentrated in the cities on the Chinese border. The region has suffered a very sharp decline in population since 1991 as Russians are moving west in search of greater economic opportunities and better living conditions, and the population is projected to drop to 4.5 million by 2015.

Water Disputes

Fairly few areas in Russia are likely to be the center of major disputes over control of fresh water resources between Russia and its neighbors. Disputes between regions within Russia are far more likely to cause trouble. The major shared water resource in Russia is the Amur River, which runs along the Sino-Russian border. The flow of the Amur is projected to increase significantly because of climate change, so conflict seems unlikely. Projected sustained water shortages in northern China could lead to interest in Lake Baikal, which holds one fifth of the world's unfrozen fresh water, more than all the North American Great Lakes combined.

Peripheral Instability

Resource and ethnic conflicts and humanitarian disasters within countries neighboring Russia are likely given their limited adaptive capacities compared to Russia. Russian humanitarian intervention appears very likely given Russia's traditionally paternalistic attitude toward its "Near Abroad." Some interventions in hard-hit regions such as Central Asia could require considerable resources.

FOREIGN POLICY IMPLICATIONS

Russia's Global Engagement

Because international engagement is not Russia's preferred option for addressing its critical interests, Russia will most likely seek autarkic solutions to problems related to climate change where it can and accept international cooperation where it must. Given the regional variation in projected climate change effects and Russia's probable expanding state capacity, Moscow will address most problems by transferring resources from one part of its territory or economy to another. Disruptions to agriculture may render the country more dependent on international markets for food and put pressure on Russia to address the ports and transportation infrastructure.

Russia is already seeking large-scale international investments in its energy sector, but as projected costs rise, Moscow will be less able to set the terms for such agreements and will have to take greater care not to alienate prospective investors. This may inhibit Russia from adventurism and "pipeline warfare" and encourage a more cooperative stance with lender countries. Russia may also have a greater need for foreign expertise in its energy industries, for example in the area of deep-water oil drilling, which could lead to more openness to foreign investment and partnerships in production.

The regional implications of climate change along Russia's periphery could easily translate into broader foreign policy concerns. Climate-induced stress and resource competition could deepen NATO and United States involvement in the Caucasus, Central Asia, and the Arctic, all areas of vital interest to Russia. Russia is unlikely to tolerate such encroachment, even if it is framed in cooperative or humanitarian terms.

Russia's Global Energy Policy

Russia's long-term financial solvency rests on its ability to ensure that it has reliable customers for its oil and natural gas exports in order to continue as a principal world energy supplier. In doing so, Russia will have to change its energy policies in reaction to the inevitable shifts in the structure of the global energy markets. Russian expectations of the future of oil prices will determine the range of climate change policy options it feels able to consider.

With the exception of natural gas through pipelines, Russia will face difficulties preserving itself as a monopolistic provider of energy because its customers will naturally seek diversification and security of supply. Russia's most important international relationships will be with its primary longer-term energy customers, particularly China.

Russia's position as a fossil fuel-based energy supplier will not be threatened if the United States and/or Europe move aggressively away from fossil fuels or away from dependence on foreign sources. Projections suggest that demand and therefore prices of Russian supply will stay high because of increases in energy demand in developing markets such as China and India. A Western move away from fossil fuels would nevertheless be troublesome and costly to Russia, forcing Moscow to reorient its exports and the attendant distribution infrastructure. Russia already faces a major medium-term shift toward Asia as its principal energy export market, but Western alternative energy policies could accelerate that shift, making more difficult the reorientation and recapitalization of Russia's energy infrastructure.

Despite projected Asian demand, Russia needs to prepare for a contingency involving a reduction in overall world demand for oil by diversifying its portfolio as an energy supplier to include more focus on natural gas and nuclear energy. In keeping with its conservative tendencies, Russia is likely to pursue both options.

Nuclear energy is a lucrative business area for Russia's energy export sector. Russia has the production capabilities and technical expertise to be a key world supplier of civilian nuclear technology and its ability to provide nuclear technology at relatively low prices makes it attractive to some customers. If Russia couples provision of nuclear reactors with reprocessing or disposal of spent fuel rods, the proliferation fears associated with customers acquiring weapons-grade nuclear materials might be assuaged. This is essentially the agreement Russia made with Iran concerning provision of nuclear fuel for the Bushehr nuclear plant.

Russia's Foreign Policy Perspectives on Climate Change

Russia does not yet consider the climate change issue a top-tier priority. Although this is likely to change over the next decade as more hard evidence of its disruptive impact surfaces, Russia still probably will feel less motivation to address climate change than the United States and others. Moscow also does have a legitimate rationale to regard climate change as not wholly negative, given that, while mainly disruptive, it is likely to generate beneficial effects in some areas. Nevertheless, as previously noted, evidence such as its recent climate change doctrine does suggest that Russia's views may be coming more into line with the international community.

Even if Russia is realizing that climate change is a serious issue, Russia's position on climate change has been, and is likely to remain, divergent from that of many other countries, including the United States. Much of the international debate on climate change mitigation, including the prospective debate at Copenhagen, has focused on measures to address the factors generating climate change, such as greenhouse gas emissions. As a primary fossil fuel supplier, Russia has little incentive to constructively engage in a debate that ultimately aims to extinguish the country's primary source of revenue. Even if it successfully manages the domestic energy challenges brought about by climate change, Russia's continued reliance on fossil fuel production will exacerbate rather than mitigate global climate change. Russia has therefore tended to treat climate change as a fait accompli and preferred to focus on adapting to and treating its effects rather than attacking the causes. Russia nonetheless is willing to accept international restrictions on emissions and initiatives toward alternative energy as long as it can be confident that demand for its oil and gas will continue to grow in Asia and the developing world.

US Engagement with Russia on Climate Change

Constructive engagement on climate change will be hampered by Russia's distrust of the United States. Beyond the legacy of decades of strategic confrontation, Russia has more proximate and concrete concerns that Western policy decisions pose direct threats to its security. Russia feels threatened by recent expansion of NATO into Eastern Europe, and particularly plans to admit more former Soviet Republic such as Ukraine or Georgia. Expanded US involvement in Central Asia, NATO operations in the Balkans, and moves toward establishing a US missile defense shield in Europe contribute to a siege mentality in the Kremlin. Russia's proactive stance in the Arctic may be an attempt to turn the tables and keep the United States and NATO at arm's length.

The Cold War experience, however, illustrates that Russia and the United States can cooperate in some areas even in an atmosphere of distrust and strategic competition. Therefore, the two may be able to cooperate and engage constructively on climate change mitigation. Moreover, Russia's assessment of the US approach to energy and climate change—in particular, how Moscow assesses US energy policy will affect oil and gas prices—will play a significant role in its internal approach to climate change. If Russia sees the United States making clear steps towards alleviating climate change, Russia will begin to adjust its own policies in response, such as by ensuring that international oil and gas prices stay high or becoming a key supplier of alternative energy technologies such as nuclear power. That said, ultimately Russia will care more about the policies adopted by Europe or China, its primary customers.

Russia is more likely to respond constructively to approaches framed in terms of energy policy rather than climate change policy. Specifically, Russia has much to gain from energy efficiency and an emphasis on technical cooperation regarding efficiency highlights action that can benefit Russia and help mitigate climate change. US companies also have expertise in areas such as deep-water drilling that might open doors to a more cooperative approach to Arctic energy exploration. In terms of alternative energy, nuclear energy may be a fruitful area for US-Russian cooperation on climate change-related issues and proliferation-resistant nuclear technology to meet the needs of international nuclear energy markets. The United States has pursued a framework for providing nuclear fuel reprocessing—a service Russia also offers—multilaterally through the Global Nuclear Energy Partnership (GNEP). Russia has expressed openness to cooperation with the United States or other countries on such approaches.

Technical and scientific cooperation on climate represents another potentially fruitful area of engagement. For example, the United States might pursue Russia's expressed interest in developing better data and models with which to assess the future effects of climate change. The United States also might involve Russia in joint efforts to model future climate scenarios. Joint modeling may not result in unified analytic results, but would at least foster a common understanding of the parameters of the issues involved. Alternatively, such an initiative might be undertaken at the non-governmental level, partnering with Russian think tanks and research institutes.

The Copenhagen Negotiations

Russia's approach to the climate change negotiations is likely to be pragmatic and opportunistic and focused on ensuring that the terms of any deal are to its benefit or at least cost Moscow nothing. Russia believes it has less at stake in the climate change debate and more freedom to extract concessions from the participants. The practical effect global climate change agreement is a secondary concern for Russia, although it would likely welcome a deal that mitigates some of the sources of climate change without constraining its oil and gas exports.

Russia's overarching strategy at Copenhagen probably will be to leave the contentious negotiations to the United States, China, India, and the European Union. Russia will look to take full advantage of the United States' interest in brokering deals with India and China, and will take every opportunity to extract favorable concessions. Moscow may position itself as an important swing player and broker between the West and the developing world.

Russia expects that the results of Copenhagen will not differ radically from the 1997 Kyoto Protocol. They presume that such a Kyoto-like system would be based on substantial individual pledges to reduce emissions according to the signatories' particular developmental situation. The terms of Kyoto were highly beneficial for Russia because emissions pledges were based on percentage changes from 1990 baseline levels: the economic collapse of the former Soviet countries after 1991 caused a major reduction in emissions, and it was easy for Russia to keep its emissions below the limitation set based on the 1990 baselines. One of Russia's objectives at Copenhagen will be to retain the favorable 1990 baseline, over European objections.

Russia's recent rhetorical recognition of the challenges posed by climate change may be a strategic move intended to better position itself to secure favorable concessions at Copenhagen. By appearing more in line with prevailing views of climate change, Russia may hope to find other participants more favorable towards its positions and to lay the groundwork to request exemption from certain climate change mitigation measures.

Russia's primary focus may be to secure favorable terms for its major capital projects and insure that any agreement reached at Copenhagen includes less restrictive terms for foreign investments. Because of its high projected investment needs, Russia will seek to secure joint international investment and credits for their energy infrastructure projects on the grounds that they will increase efficiency and thereby help mitigate climate change. Gazprom, for example, argues that investment to help them upgrade their gas pipeline system and credits for reducing gas flaring by a significant amount would be beneficial in combating climate change. Gazprom and other elements of the Russian energy industry have identified climate-related international investment as a major potential source of funding.

Although Russia may exhibit a flexible position on climate change, it has very little room to maneuver in terms of concessions that would negatively affect its oil and gas sector. It is also very unlikely to accept conditions that it regards as threatening its sovereignty. Russian willingness to accept the terms of an international climate deal could evaporate if these core interests are challenged or its paranoia over an international anti-Russian conspiracy is triggered.

End Notes

[1] This section is extracted from the Executive Summary of the Phase I report (see Scope Note): ***Russia: Impact of Climate Change to 2030, A Commissioned Research Report*** (NIC 2009-04, April 2009). Some of the judgments in this chapter (Phase II) may differ from the Phase I report.

[2] See the Phase I report, p. 35 for a discussion of adaptation and resilience capacity.

[3] Data as of 2005; see Migration Policy Institute: http://www.migrationinformation.org/DataHub/charts/6.1.shtml; viewed 23 Sep 2009

[4] Andrew C. Kuchins, *Alternative Futures of Russia to 2017* (Washington DC: Center for Strategic and International Studies, December 2007).

[5] Although Russia hosts nearly 160 ethnic groups, the only minorities with over a 1 percent representation in the population are Tatars, Ukrainians, and Chuvash.

[6] World Bank and International Finance Corporation, *Energy Efficiency in Russia: Untapped Reserves* (Washington, 2008)

[7] Заседание Совета Безопасности Российской Федерации, (Russian Federation Security Council Meeting) «О защите национальных интересов Российской Федерации в Арктике» (On the defense of Russian Federation's national interests in the Arctic), 17 September 2008; NIC 2009-04, p. 34.

INDEX

#

20th century, 7, 9
21st century, 11, 13

A

access, 34, 35, 36, 60, 66, 91, 92, 94
activism, 72, 86
adaptation, 25, 31, 34, 35, 40, 47, 48, 59, 104
adaptations, 28
additives, 41, 68
adverse conditions, 96
Africa, 1, 57
age, 5, 60, 69, 96
agencies, 6, 76, 82
agriculture, 4, 5, 26, 27, 40, 41, 60, 62, 68, 74, 80, 97, 99
air quality, 35
air temperature, 6, 7, 8, 11, 12, 16, 43
alternative energy, 80, 87, 100, 101, 102
Amur River, 97, 98
Aral Sea, 96
Armenia, 95, 97
Asia, 1, 4, 10, 12, 13, 26, 29, 32, 33, 43, 44, 45, 46, 57, 59, 60, 62, 64, 68, 70, 85, 92, 94, 95, 96, 98, 99, 100, 101
assessment, 2, 6, 12, 13, 18, 31, 47, 101
assets, 8
atmosphere, 7, 10, 101
ATO, 101
attribution, 35
authorities, 27, 71, 72, 76, 79, 86
authority, 34, 71, 72, 77, 83, 85
awareness, 73, 78
Azerbaijan, 95, 97

B

balance of payments, 41
Balkans, 92, 101
Barents Sea, 13, 24, 41, 88, 90, 95
base, 7, 30, 32, 34, 36, 93
Belarus, 92
beneficial effect, 100
benefits, 18, 28, 35, 60, 79
Bengal, Bay of, 43
bias, 43, 44, 46
biodiversity, 70
birth rate, 36, 69, 70
black market, 72
Black Sea, 5, 32, 63, 92
boundary conditions, 9
bridges, 67, 88, 91
bureaucracy, 82

C

calculus, 40
campaigns, 71
candidates, 71
capital expenditure, 67, 87
capital projects, 103
carbon, 48, 79
cargoes, 32
Caribbean, 1, 57

Caspian Sea, 71
Caucasus, 4, 5, 13, 22, 26, 30, 33, 42, 60, 62, 63, 68, 70, 76, 93, 95, 97, 99
Central Asia, 4, 26, 33, 43, 46, 62, 64, 68, 70, 78, 95, 96, 97, 98, 99, 101
Central Europe, 92
central planning, 64
challenges, 3, 4, 9, 21, 22, 25, 26, 27, 28, 29, 31, 33, 58, 59, 60, 61, 62, 63, 64, 66, 67, 69, 72, 73, 75, 78, 79, 81, 83, 84, 85, 90, 91, 101, 103
chemical, 9, 41, 68
chemical interaction, 9
China, 1, 4, 33, 36, 38, 40, 44, 57, 60, 62, 63, 64, 70, 89, 92, 94, 96, 97, 98, 99, 101, 102
CIA, 48, 54, 55, 58
circulation, 8, 10, 11, 71
cities, 4, 19, 31, 34, 42, 52, 60, 62, 69, 71, 73, 74, 75, 86, 90, 97
citizens, 59
civil society, 71, 72
Climate Change Science Program, 47
climates, 47
climatic shifts, 75, 76, 79
clustering, 97
CO2, 10, 11
Coast Guard, 30, 94
coercion, 84
Cold War, 33, 101
collaboration, 31, 73
color, iv
commerce, 67
commercial, 40
commodity, 19, 40, 67, 72, 87
communities, 8, 63, 86
community, 2, 58, 100
comparative advantage, 87
competition, 60, 76, 78, 80, 95, 99, 101
complications, 24, 79
composition, 8, 9
conditioning, 29, 75
conference, 1, 57
Conference Report, 57
conflict, 74, 76, 77, 83, 93, 95, 98
conflict of interest, 83
confrontation, 101
conspiracy, 103
construction, 3, 24, 25, 33, 61, 66, 88
consumers, 24
consumption, 20, 48, 65, 87
contamination, 69, 96
contingency, 100
controversial, 18, 23
convergence, 38
cooperation, 81, 93, 98, 102
cost, 21, 41, 59, 67, 88, 102
cotton, 26, 68
creep, 67
crimes, 34
critics, 80
crop, 35, 60, 62, 68, 76
crops, 3, 16, 26, 27, 41, 62, 68
crude oil, 32
cultivation, 68
currency, 67
customers, 24, 66, 85, 89, 92, 99, 100, 101
cycles, 43, 77

D

Dagestan, 77
danger, 75
data collection, 83
datasets, 12
decentralization, 85
deficiencies, 2, 64, 75
deforestation, 48
degradation, 9, 77, 86, 88
Denmark, 93
Department of Agriculture, 40
Department of Homeland Security, 30
dependency ratio, 36, 38
deposition, 9, 35
deposits, 14, 65, 66, 67, 94
depth, 7, 10, 11
destination, 4, 33
destruction, 77
developing countries, vii, 3, 61
disaster, 31, 82
discharges, 9, 17

discrimination, 76
diseases, 27, 30, 71
dislocation, 76
distribution, 8, 21, 27, 34, 44, 45, 75, 100
district heating, 52
diversification, 38, 40, 70, 87, 99
diversity, 79
division of labor, 81
draft, 53
drainage, 5
drawing, 2
drinking water, 36
drought, vii, 2, 27, 32, 61, 76
drying, 63, 76, 90, 96

E

earnings, 23
East Asia, 43, 44, 60, 92, 94
Eastern Europe, 101
ecological systems, 47
ecology, 70, 86
economic activity, 67, 79, 80, 85
economic crisis, 95
economic development, 82
economic downturn, 67, 95
economic efficiency, 80
economic growth, viii, 38, 58, 64, 94
economic incentives, 88, 96
economic migrants, 97
economic performance, 22, 83
economic power, 23, 66, 85
economic relations, 72
economic resources, 35
economic well-being, 35
economics, 65, 80
economy, 19, 22, 32, 38, 59, 64, 67, 74, 80, 86, 87, 90, 99
ecosystem, 35
education, 34, 40, 41, 69
educational system, 69
Efficiency, 104
elderly population, 74
election, 71
electricity, 21, 22, 75, 89, 93
emergency, 59, 82, 94
emergency response, 59, 82
employment, 35, 60, 95
employment opportunities, 95
encephalitis, 30
energy, viii, 3, 19, 20, 21, 22, 40, 48, 58, 59, 60, 61, 64, 65, 66, 67, 69, 73, 77, 79, 80, 85, 86, 87, 88, 89, 91, 92, 93, 94, 99, 100, 101, 103
energy consumption, 20, 65, 87
energy efficiency, 80, 87, 102
energy markets, 99, 102
energy prices, 22, 40, 87
energy supply, 21, 22
engineering, 90
environment, 26, 38, 73, 91
environmental change, 67
environmental degradation, 77, 86, 88
environmental effects, 73
environmental harm, 83
environmental impact, 40
environmental issues, 59, 73, 77
Environmental Protection Agency, 14
environmental resources, 34
equity, 35, 36
erosion, 5, 13, 18, 28
ethnic diversity, 79
ethnic groups, 70, 76, 104
ethnic violence, 60
EU, 55, 92, 94
Eurasia, 14, 15
Europe, 5, 17, 29, 32, 69, 89, 92, 94, 96, 99, 101
European market, 23
European Union, 60, 92, 102
evaporation, 13
evidence, 18, 71, 100
expenditures, 81, 87
experiences, 5
expertise, 82, 90, 99, 100, 102
exploitation, 65, 86, 93, 94
exploration, 22, 102
export market, 100
exporter, 89
exports, 19, 22, 24, 65, 87, 89, 92, 93, 99, 102

exposure, 35
external shocks, 90
extraction, 25, 66, 74, 82, 93
extremists, 84

F

farmers, 27, 41
fears, 70, 100
Federal Emergency Management Agency, 30, 82
Federal Security Service, 83
feedback, 15, 44
FEMA, 30, 82
fever, 29, 30
financial, 19, 40, 53, 86, 89, 99
financial crisis, 19, 53
first responders, 59, 85
fish, 71, 95
fishing, 71, 74, 77, 95
flexibility, 90
flooding, vii, 2, 17, 18, 19, 29, 30, 31, 34, 42, 61, 63, 64, 67, 68, 69, 75
floods, 3, 16, 17, 32, 62, 70, 73, 82
flora, 70
flora and fauna, 70
food, vii, 2, 25, 26, 27, 34, 35, 36, 42, 54, 56, 61, 73, 75, 76, 77, 80, 99
food chain, 77
food prices, 76
food production, 34, 35
food security, 36
force, 1, 22, 58, 66, 76
forecasting, 31, 59
foreign exchange, 87
foreign investment, 99, 103
foreign policy, 65, 99
forest fire, 17, 30, 70, 77, 82
formation, 14
freshwater, 15
frost, 5, 7, 16, 26
fruits, 26, 68
FSB, 83, 96
funding, 89, 103

G

GDP, 35, 36, 38, 71
GDP per capita, 38
Georgia, 95, 97, 101
Germany, 92
global climate change, 101, 102
global security, 89
global warming, 9, 44
goods and services, 89
governance, 71, 79
governments, 59, 65, 82, 84, 85, 86
Great Lakes, 98
greenhouse, 6, 9, 10, 11, 21, 38, 100
greenhouse gas emissions, 11, 21, 38, 100
greenhouse gases, 6, 9
grids, 22
groundwater, 14
grouping, 40
growth, 10, 24, 38, 39, 40, 64, 65, 81, 83, 94
guidelines, 79

H

Haiti, 36
hard currency, 67
hazards, 17, 28, 29
health, 30, 35, 41, 69, 70, 73, 75, 86
health problems, 70
history, 6, 67
host, 26, 29, 31, 72, 78
hostility, 78
housing, 29
human, viii, 6, 18, 22, 27, 29, 34, 35, 36, 38, 48, 58, 64, 69, 70, 77
human actions, 70
human capital, viii, 34, 35, 38, 58, 64, 69
Human Development Report, 49, 51, 55
human health, 29, 36
human resources, 36, 69
human welfare, 38
hydrocarbon deposits, 65
hydrocarbons, 40, 87
hydroelectric power, 3, 21, 61

I

ice, 3, 5, 6, 7, 8, 10, 11, 15, 17, 24, 28, 29, 32, 54, 60, 62, 68, 90, 91, 93, 94, 95
ice pack, 60
Iceland, 33
identity, 72
immigrants, 4, 62, 78
immigration, 32, 70, 96
immune system, 75
Impact Assessment, 54, 55
impacts, vii, viii, 2, 3, 4, 6, 10, 12, 14, 19, 21, 22, 25, 27, 28, 30, 31, 34, 38, 39, 41, 42, 47, 48, 49, 50, 51, 52, 58, 59, 61, 62, 63, 65, 67, 68, 70, 72, 73, 76, 77, 78, 79, 85, 97
imports, 93
improvements, 44, 91, 92, 97
incidence, vii, 2, 22, 24, 61, 68, 71
income, 36, 71
independence, 81
India, 1, 36, 43, 44, 57, 60, 89, 99, 102
indigenous peoples, 70, 74
individuals, 72
Indonesia, 36, 45
industrialized countries, vii, 3, 59, 61, 67
industries, 19, 22, 65, 67, 79, 85, 87, 89, 99
industry, 3, 22, 23, 24, 40, 53, 61, 65, 71, 89, 92, 97, 103
infrastructure, viii, 2, 3, 19, 31, 36, 41, 42, 58, 59, 60, 61, 63, 64, 65, 66, 67, 68, 73, 74, 78, 79, 81, 82, 83, 85, 87, 88, 89, 91, 94, 99, 100, 103
insects, 17, 71
institutions, 34, 40, 75
insulation, 23
integration, 86
integrity, 23, 67
interdependence, 81
interest groups, 59, 77
international investment, 99, 103
international relations, 99
investment, 85, 88, 89, 91, 92, 99, 103
investments, 59, 67, 79, 82, 88, 91, 92, 99, 103
investors, 88, 99
Iran, 89, 96, 100
irrigation, 4, 27, 40, 41, 60, 62, 68, 74, 90
IRS, 7
islands, 6, 17, 64
issues, 27, 47, 59, 72, 73, 77, 86, 88, 102

J

Japan, 44, 85
jurisdiction, 7, 93

K

Kabardino-Balkaria, 77
Kazakhstan, 96
Kenya, 76
Kola Peninsula, 24
Korea, 44

L

labor market, 78
labor shortage, 97
lakes, 5, 18, 71
landscape, 5, 18
lead, viii, 2, 3, 10, 15, 17, 20, 22, 28, 29, 31, 58, 61, 68, 69, 70, 75, 78, 82, 91, 98, 99
leadership, 59, 81
legislation, 79
level of education, 34
life expectancy, 36, 69
light, 21, 46, 66, 75
liquefied natural gas, 87, 89
literacy, 36, 38
living conditions, 98
local authorities, 72, 86
local conditions, 40
local government, 65, 85

M

magnitude, 66, 76
majority, 11, 74, 93, 95, 96
malaria, 30
management, 27, 82
mass, 15, 31, 74

measurements, 8, 17, 73
media, 48, 49, 50, 51, 52, 53, 54, 95
median, 13, 43, 46
melt, 7, 18, 21, 22, 28, 29, 32
melting, vii, 2, 15, 24, 59, 61, 63, 64, 65, 66, 67, 70, 88, 90, 91, 93, 94, 95
Mendeleev, 93
methodology, 38
Mexico, 1, 36, 57
middle class, 38
migrants, 33, 60, 78, 95, 96, 97
migration, 4, 16, 33, 42, 43, 60, 62, 63, 64, 68, 69, 74, 77, 78, 95, 96, 97
military, 32, 33, 80, 94, 97
minorities, 76, 78, 104
minority groups, 76, 77
modeling, 9, 10, 11, 38, 60, 63, 102
models, 6, 8, 9, 10, 11, 12, 13, 14, 15, 42, 43, 44, 45, 46, 48, 74, 102
moisture, 11
Mongolia, 4, 33, 62, 95, 96
monitoring, 31, 86, 94
monopoly, 28, 89, 90
mortality rate, 69
Moscow, 5, 8, 11, 19, 20, 26, 30, 34, 48, 49, 50, 51, 52, 53, 54, 55, 60, 65, 67, 69, 71, 72, 73, 75, 78, 79, 80, 81, 85, 87, 98, 99, 100, 101, 102
MPI, 45
MRI, 45
Muslims, 70

N

NAS, 10, 12, 13, 43
national interests, 104
national parks, 86
national security, 1, 2, 22, 57, 58, 81, 82
NATO, 33, 55, 94, 99, 101
natural disaster, 31
natural disasters, 31
natural gas, 41, 87, 89, 90, 92, 99, 100
natural resource management, 80
natural resources, 93
negative effects, vii, 2, 59, 60, 61
North Africa, 1, 57
North America, 29, 32, 94, 98
North Caucasus, 4, 22, 42, 62, 63, 68, 76, 97
Northeast Asia, 85
Norway, 94, 95
nuclear energy, 100, 102
nuclear power, 22, 87, 89, 101
nutrients, 14

O

oceans, 8, 45
officials, 76, 79, 93
oil, 3, 22, 23, 32, 40, 41, 59, 61, 65, 66, 67, 71, 77, 78, 79, 83, 86, 87, 88, 89, 91, 92, 93, 94, 99, 100, 101, 102, 103
oil spill, 88
openness, 81, 99, 102
operations, 22, 25, 28, 101
opportunities, 21, 33, 34, 88, 95, 98

P

Pacific, 1, 2, 5, 10, 18, 25, 29, 32, 57, 86, 94
parallel, 92
paranoia, 80, 103
parity, 87
participants, 58, 102, 103
per capita income, 71
performance, 45, 47
pests, 27, 29, 41, 48, 68, 71, 97
petroleum, 3, 19, 21, 22, 23, 28, 32, 61
petroleum industry, 3, 22, 23, 61
Philippines, 36
pipelines, 23, 24, 59, 67, 73, 88, 89, 90, 91, 92, 99
plant diseases, 68
plant growth, 14
plants, 23, 35
pluralism, 80
Poland, 92
polar, 5, 32, 33, 94
policy, 2, 31, 58, 65, 79, 81, 83, 85, 87, 93, 99, 101
policy options, 99
policy reform, 31
policy responses, 85

policymakers, 31
political crisis, 80
political leaders, 33
political party, 72
political power, 75
political system, 83
population, 5, 10, 29, 32, 34, 35, 36, 38, 40, 59, 60, 64, 69, 70, 71, 74, 75, 76, 78, 85, 96, 97, 98, 104
population density, 36, 38
population growth, 38
portfolio, 100
precedent, 85, 96
precipitation, vii, 2, 3, 5, 6, 7, 8, 11, 12, 13, 14, 15, 21, 26, 41, 43, 44, 45, 46, 47, 61, 62
preparation, iv, 79
preservation, 73, 86
President, 40, 92
prevention, 30, 82
production costs, 87
production targets, 24
productivity, 21, 35, 47, 69, 73, 74
project, 10, 15, 17, 25, 88, 90
proliferation, 89, 100, 102
property rights, 34
protected areas, 86
protection, 30, 86
public awareness, 73
public health, 30, 69, 75
publishing, 86

Q

quality of life, vii, 2, 61
quartile, 13, 36, 46

R

rain forest, 5
rainfall, 10, 13, 43, 44, 45
recognition, 93, 103
recommendations, iv
redistribution, 80, 90, 96
redundancy, 81
reform, 83
Reform, 53
reforms, 31
refugees, 78, 95
regional problem, 85
regions of the world, 1, 57
replacement, 41, 59, 66, 67, 88
repression, 59, 72
reprocessing, 100, 102
reputation, 82
requirements, 19, 21, 24, 27, 28, 31
reserves, 67, 81, 86, 87, 93
resilience, vii, 3, 35, 37, 40, 47, 80, 81, 97, 104
resistance, 81
resolution, 9, 10, 43, 44, 45
resource management, 80
resources, 25, 31, 34, 35, 36, 41, 59, 60, 65, 69, 74, 75, 76, 78, 80, 85, 86, 91, 92, 93, 94, 95, 98
response, 7, 9, 11, 30, 44, 59, 61, 66, 74, 75, 78, 79, 82, 86, 101
response capacity, 86
restrictions, 101
restructuring, 71
revenue, 3, 61, 65, 87, 101
rhetoric, 80, 87
rights, iv, 34, 72, 95
risk, 2, 22, 23, 29, 30, 31, 32, 33, 34, 36, 41, 64
risks, 3, 23, 28, 31, 32, 41, 55, 61, 63, 64, 68, 88, 93
river basins, 17, 67
river flows, 28, 63, 65, 67, 68, 71, 89
river systems, 28
routes, 60, 67, 78, 92, 94
runoff, 9, 14, 15, 16
rural areas, 19, 73, 74, 75, 78
rural population, 40, 74

S

saturation, 96
savings, 21, 29, 52
Scandinavia, 93
school, 71, 95
sea level, 2, 18, 31, 32, 36

sea-level, 31, 41, 63, 71
sea-level rise, 31, 41, 63, 71
seasonality, 20, 45
security, 1, 2, 22, 33, 34, 55, 57, 58, 59, 73, 80, 81, 82, 83, 88, 92, 99, 101
sensitivity, 10, 17, 36, 37, 39, 56, 80
services, iv, 19, 59, 89
settlements, 88
severe stress, 60
ships, 55, 94
Siberia, 3, 6, 7, 9, 11, 12, 13, 15, 16, 23, 24, 26, 27, 28, 30, 62, 64, 67, 68, 71, 73, 74, 88, 90, 94, 95
side effects, 81
simulation, 9, 12, 44, 45, 47
simulations, 14, 44, 45, 47
social capital, 34
social group, 34, 40
social network, 72, 74, 96
social organization, 72
social problems, 78
society, 27, 34, 36, 40, 59, 71, 72, 75, 84
solidarity, 49, 51, 55
South Asia, 43, 44, 96
Southeast Asia, 1, 43, 45, 57
sovereignty, 72, 86, 103
Soviet Union, 7, 69, 81, 83
specialization, 81
species, 2, 16, 70
spine, 64
St. Petersburg, 6, 17, 31, 32, 63, 67, 75, 78
stability, 75, 81, 83
stabilization, 10
standard of living, 19
state, viii, 2, 3, 22, 28, 35, 36, 58, 59, 61, 64, 65, 71, 72, 73, 74, 75, 76, 77, 78, 79, 80, 81, 82, 83, 84, 86, 88, 89, 90, 91, 94, 97, 98
state control, 86
state intervention, 81
states, 11, 32, 33, 37, 80, 92, 93, 94, 95, 96, 97
storms, 24, 28, 29, 42, 68, 69, 82, 94
streams, 14, 71
stress, viii, 4, 35, 42, 54, 56, 58, 60, 62, 63, 68, 69, 70, 72, 74, 75, 76, 91, 96, 99
stretching, 28
structure, 34, 99
subsistence, 74, 76, 86
sulfur, 35, 36
sulfur dioxide, 36
supplier, 89, 92, 99, 100, 101
supply chain, 27
surging, 41
survival, 19, 28
sustainable development, 83
sustainable economic growth, viii, 58, 64
syndrome, 29
synthesis, 47

T

techniques, 23, 24, 40, 66
technological change, 10, 40, 64
technological progress, 38
technologies, 101
technology, 35, 40, 100, 102
temperature, 5, 6, 7, 8, 9, 11, 12, 13, 20, 21, 26, 43, 44, 45, 46, 60, 73
tension, 72, 95, 97
tensions, 2, 4, 42, 58, 60, 63, 76, 77, 78, 97
territory, 3, 18, 23, 33, 60, 62, 63, 64, 65, 67, 71, 74, 93, 97, 98
threats, 42, 67, 85, 101
Tibet, 43, 46, 47
time frame, 2, 58
traditional practices, 75
transmission, 21, 22, 93
transport, 9, 28, 66, 67, 80, 94
transportation, viii, 21, 22, 23, 28, 29, 32, 58, 59, 60, 63, 65, 66, 67, 68, 75, 76, 91, 99
transportation infrastructure, viii, 58, 59, 60, 63, 66, 67, 99
treatment, 32, 69, 75
tsunami, 18
tundra, 5, 8, 70, 97
Turkey, 92

U

Ukraine, 32, 74, 92, 93, 101
UN, 21
unemployment rate, 63
United, 4, 5, 33, 49, 51, 52, 55, 60, 62, 71, 94, 99, 100, 101, 102
United Nations, 49, 51, 52, 55
United States, 4, 5, 33, 60, 62, 71, 94, 99, 100, 101, 102
urban, 31, 32, 40, 60, 63, 64, 69, 74, 75, 77, 78
urban area, 40, 69, 77
urban areas, 40, 69, 77
urban population, 69
USSR, 40
Uzbekistan, 96

V

valleys, 43
variables, 35, 40, 51
variations, 44, 45
varieties, 3, 26, 27, 41
violence, 60, 76, 78
volatility, 22, 63, 90
vulnerability, 23, 28, 31, 36, 38, 41, 47, 80, 87

W

Washington, 53, 55, 104
waste, 75, 88
water, vii, 2, 3, 4, 7, 8, 9, 14, 17, 21, 23, 24, 25, 27, 28, 30, 32, 36, 39, 47, 60, 61, 62, 63, 68, 69, 71, 74, 75, 76, 90, 91, 94, 96, 98, 99, 102
water quality, 30
water resources, 36, 98
water shortages, 3, 4, 25, 60, 62, 63, 68, 69, 76, 90, 96, 98
weakness, 38, 40
wealth, 19, 24, 38, 39, 40, 65, 83
welfare, 36, 38, 87
Western Siberia, 13, 15, 66
workers, 20, 33
World Bank, 87, 104
worst case scenarios, 80

X

xenophobia, 60, 78

Y

yield, 16, 51

Z

Zone 1, 20
Zone 2, 20
Zone 3, 20